D0207984

VALUE-ADDED SERVICES FOR NEXT GENERATION NETWORKS

VALUE-ADDED SERVICES FOR NEXT GENERATION NETWORKS

Thierry Van de Velde

 Auerbach Publications
Taylor & Francis Group
New York   London

CRC Press is an imprint of the
Taylor & Francis Group, an **informa** business

Auerbach Publications
Taylor & Francis Group
6000 Broken Sound Parkway NW, Suite 300
Boca Raton, FL 33487-2742

© 2008 by Taylor & Francis Group, LLC
Auerbach is an imprint of Taylor & Francis Group, an Informa business

Library of Congress Cataloging-in-Publication Data

Velde, Thierry Van de.
    Value added services for next generation networks / Thierry Van de Velde.
       p. cm. -- (Informa telecoms & media ; no. 7)
    Includes bibliographical references and index.
    ISBN 978-0-8493-7318-3 (alk. paper)
    1. Telecommunication. 2. Customer services. 3. Value added. I. Title.

TK5101.V45 2007
621.382--dc22
                                       2007017685

Visit the Taylor & Francis Web site at
http://www.taylorandfrancis.com

and the Auerbach Web site at
http://www.auerbach-publications.com

Contents

Chapter 1

Basic Forces

Welcome to this first chapter!

In this book, let's undertake a quest for the real "added value" of modern (tele)communication services.

I never liked general statements such as "efficient communications increase productivity" or "social persons communicate frequently." Aren't our buddies and colleagues being flooded by the numerous calls, messages and streaming media? What are we, our employers or advertisers willing to pay for true communication means? Are our children going to adopt them or will they invent their own? Will our machines start communicating in unpredictable ways?

This book is not only about technology itself; also on how it is being used and could be used in the best possible way.

We have a long way to go to be able to conceive Value-Added Services (VAS) for Next Generation Networks (NGN) — at least that's my feeling at the start of this book. A NGN is not so easy to define either.

It's going to be a dangerous journey; we might lose ourselves in the wilderness of networks, the ecosystem of companies, or daunting traffic predictions. We might end up with a list of dozens of possible services with no way to prioritize or even understand their extent.

Let's therefore weapon ourselves for the trip!

In this chapter, we'll take a helicopter tour over the basic forces ("dimensions"?) which to my opinion are structuring modern communication means:

- Means and appropriateness
- Contextual and interactive portions
- Device, connectivity and service

1

- Cellular, mobile and wireless
- Presence and availability
- Immediate and deferred mode
- On-demand, triggered and periodic
- Personal contacts and tribes
- Location based

Traditionally, communication means would have been classified as

- Fixed or mobile
- Voice, text, video or data
- Pre- or postpaid
- Retail or wholesale
- Circuit- or packet-switched
- Point-to-point or multiparty
- Best-effort or guaranteed Quality-of-Service
- Private or public

But in the context of the Next Generation Networks, these refer to the properties of bit pipes (connectivity) rather than to recognizable communication services.

And this book is about genuine communication services, not about connectivity.

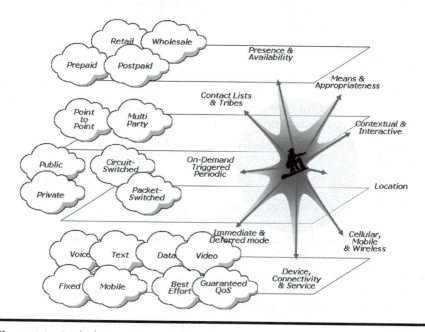

Figure 1.1   Basic forces structuring modern communication means.

Let's therefore try and walk a different path; let's try and uncover the essence of a communication service.

But before that, perhaps we can start by defining and delineating communication against its two natural rivals: storage and computing. The relationship between communication, storage and computing is in constant evolution; allied to provide one communication service, they are enemies for the next one.

1.1 Symbolic Devices

We all know, from what we experience and within ourselves, that our conscious acts spring from our desires and our fears. Intuition tells us that that is true also of our fellows and of the higher animals. We all try to escape pain and death, while we seek what is pleasant. We are all ruled in what we do by impulses; and these impulses are so organized that our actions in general serve for our self preservation and that of

Figure 1.2   Albert Einstein playing the violin.

the race. Hunger, love, pain, fear are some of those inner forces which rule the individual's instinct for self preservation. At the same time, as social beings, we are moved in the relations with our fellow beings by such feelings as sympathy, pride, hate, need for power, pity, and so on. All these primary impulses, not easily described in words, are the springs of man's actions. All such action would cease if those powerful elemental forces were to cease stirring within us. Though our conduct seems so very different from that of the higher animals, the primary instincts are much alike in them and in us.

The most evident difference springs from the important part which is played in man by a relatively strong power of imagination and by the capacity to think, aided as it is by language and other symbolic devices.

Thought is the organizing factor in man, intersected between the causal primary instincts and the resulting actions. In that way imagination and intelligence enter into our existence in the part of servants of the primary instincts. But their intervention makes our acts to serve ever less merely the immediate claims of our instincts.

—Albert Einstein

Following the reasoning of the great scientist, language and symbolic devices help us to construct imagination and intelligence: the tools that mitigate the effect of our primary instincts on our acts.

From the picture above, I'm sure Einstein would have named music as one of the symbolic devices.

Some symbolic devices such as the mathematical and physical formalisms have helped mankind to construct the intelligence to address basic impulses such as cold (with electricity from nuclear energy), fear (with insight on the origin and evolution of the universe) and need for power (with atomic weapons).

Science, art, medicine, religions, political systems, law and economics have become the major symbolic devices in today's world.

It is in that prestigious enumeration that we would like to add telecommunications one day: not what is said in the billions of daily phone calls, messages, and Web pages; not the copper wire, glass fiber, and ethereal networks themselves; but the set of underlying protocols, network structures, and procedures used by mankind to communicate.

Communication is essential for life — would the Darwinist evolution have taken place without it?

But in the big picture of things, speech and languages are relative novelties.

Early humans must have developed speech and languages for rather functional reasons, such as to give each other simple orders, to teach children faster how to survive, or to join forces when hunting. Speech and languages were born as the direct servants of the primary instincts.

As cognitive intelligence grew, so did the number of abstract topics (e.g., love, war, religion, possessions, tribes) and the number of words available to construct spoken phrases. Speech and languages became tools to construct intelligence and imagination, through a process of sustained communication.

It is indeed hard to imagine building intelligence and imagination in complete isolation. As the great Einstein would say, "The secret of creativity is knowing how to hide your sources."

The relationship between communication, intelligence and imagination is fascinating. Our knowledge economy depends on people finding the right balance. What will be the future role of communication?

1.2　Communication and Storage

Did communication grow out of knowledge or the reverse?

The human need for communication fuelled the development of different means which today we would call "media." In order of appearance: speech, theatre, the telegraph, telephone, radio, TV, cellular communications, sms, chatting, blogging, podcasting, and so on.

Most of these communication channels could not have emerged without an underlying support at the end points (some "storage" means): more brain cells for speech, the grotto wall for paintings, papyrus for writing, printed books, newspapers, photo paper, movie film, vinyl records, magnetic storage, computer memory, file systems, optical storage (CD, DVD), nonvolatile memory, and so on.

These storage means are the support for all our knowledge. The rest are thoughts, experience and imagination.

In the course of history, communication and storage have gone hand in hand. More brain cells meant more speech. More writing skills meant an increased need for papyrus. Gutenberg's printed bible was read by a whole new category of readers, who couldn't have afforded the medieval calligraphy. And so on.

But today, communication and storage seem to compete with each other on a daily basis, and to *exclude* each other for various reasons.

Should we have a conference call on this project (communication) or await everyone's written input (storage)? Knowing that most conference calls are not minuted (stored) anymore …

Should I e-mail this document to my colleagues (communication) or post it on our intranet (storage)?

Am I going to take incoming calls (communication), or are the callers going to leave a voicemail (storage)?

Is attending that Webinar (communication) really going to be more useful than using a search engine on the same topic (storage)?

In my instant messenger, do I say hello (communication) or do I just publish my presence (storage)?

Is digital TV a truly interactive experience with potentially thousands of channels on very specialized and local topics (communication), or will it only allow me to watch previous shows and download movies on demand (storage)?

Why is wireless broadband access (WiFi) suddenly built into this new portable gaming device? For multiplayer games (communication) or to download new games (storage)?

The line may seem thin between "genuine" communication and "just" the electronic transmission of stored media. But it isn't!

Genuine communication differentiates itself from stored media transmission with *timeliness, expectation,* and *unpredictability.*

As the time increases between the production of the data and their transfer, such as in the case of books, art exhibitions or song downloads, people tend not to qualify this as communication anymore. Even though press critics clearly provide feedback on books, movies and recorded music (the bidirectional aspect).

If the initiator expects a response, communication takes place, even if technically no response may actually come. Sending an application letter to a company would qualify as genuine communication, posting your curriculum vitae online wouldn't.

Unpredictability is required too. Is *using a search engine* genuine communication? No — do the same query in 5 minutes and it will result in the same results. In a corporate organization, the outcome of a Web training is less predictable (in terms of questions from the attendants, unexpected disagreement, next actions arising, etc.) than the outcome of a written note on the same subject, to the same list of persons (of which the impact will probably be more limited). The Web training would qualify as genuine communication; the note would be only transmission of stored media.

The volume of data, or the pace at which it is transferred, plays no role to qualify an interaction as "communication." For example, one sigh says more (can better be qualified as genuine communication) than a thousand words. And people are willing to pay 50 cents to send one byte of information (a single digit) in a televoting sms taking 10 seconds to reach the TV show. But 1 Euro sounds expensive for a megabyte of high-speed mobile data traffic (for example, containing 10 e-mails). A higher volume and speed are not necessarily perceived as more valuable communication. There's a common belief that broadband services actually kill more value (the revenue of narrowband services) than the amount of new value they create. Volume and bandwidth are rapidly commoditized, genuine communication services probably won't be.

Is genuine communication more attractive and valuable than long-term storage?

Not necessarily. Things change with age. Teenagers may have many contacts with peers, but as life advances, elderly people love their music collection, movies, books and photo albums.

For sure, the transfer of stored media over IP networks has become a commodity, and it is still a question whether the residential end user is willing to pay for some quality of service, even if enforced end-to-end — even if corporations, large and small, might find telepresence (videoconferencing over IP) extremely valuable.

It is essential for modern telecommunication service providers to position themselves relative to genuine communication versus the transmission of stored media: should they invent and launch new communication means, or provide high quality bit pipes for stored contents they don't interpret? Are they going to deliver Value-Added Services or just Next Generation Networks? *Are they going to be the diplomat or the pigeon?*

Failing to specialize and be successful in the one or the other, some of them may be tempted to become *the bard* — the creators of stored content — resulting in investments in portals to *walled gardens*, movie-on-demand servers, soccer broadcasting rights, music catalogs, and so on.

These digressions from the original company mission and core business cannot be justified with the need to demonstrate the usefulness of the bit pipes. Remember these nineteenth-century railway-owned station buffets — passengers arrived there by very useful trains (the bit pipes) but who would have considered them to be representative for the visited town's culinary scene (the content)? Railway companies don't attract the best cooks or the most motivated waiters.

In conclusion, it sounds fair to say that storage (including the commoditized transmission of stored media) on one side, and genuine communication means on the other, have today become enemies, and can tear apart each other's business models, companies and end users.

1.3   Communication and Computing

True communications are clearly fuelled by the advances in computing.

It sounds unbelievable that the major inventions for the digital age (the transistor, the microchip, magnetic and optical storage) all happened less than 50 years ago, that the Internet appeared less than 20 years ago. That the GSM handsets evolved from a cell phone to a networked personal computer in less than 10 years — at the center of their own personal access network of Bluetooth headsets, car kits and keyboards.

Computing is still in its expansion phase and the movement seems to accelerate.

Wireless broadband technology is going to expand the reach of the Internet into remote areas without copper pairs or even without electricity; therefore creating new markets for low cost computing devices (laptop PCs, batteries and solar panels).

The expansion of computing is not restricted by the number of humans — therefore by the human assimilation bandwidth or human desire to communicate. Cars and household appliances contain computers. Our electricity meters might soon contain one. Radio-frequency identification (RFID) tags would allow every item in a clothes shop to contain a disposable computer. Watches will soon contain a computer transmitting the heartbeat rate to the Internet, via Wibree and your mobile phone, during your jogging. A personal trainer might even return advice to you, in real time, to your Bluetooth headset, again containing a computer.

As computing expands toward the edges, so will transmission and storage, but will the telecommunication sector (i.e., telecommunication service providers) benefit from this movement?

Also, what are the forces limiting this expansion?

Some people feel unhappy about the inefficient use of computing resources made by modern operating systems and distributed software. It is not sure that people will keep buying ever more computing power to cope with this. Virtualization, Web services and Display-over-IP (DoIP) terminals will save computing resources, by simplifying the user device and recentralizing computing, storage and software as a pool of on-demand resources.

The success of memory sticks and media players using media compression technology (e.g., MP3, AAC, MPEG4) resulted from the need to save on nonvolatile memory but also illustrates a trend away from computing power. Today's consumers appear to be willing to pay the same price for a high-end portable media player as for a basic desktop computer.

Can true, valuable communication appear between computers, rather than the mere transmission of stored media? We are not so far from this. Computers have started relaying packets for peer-to-peer services, without their human owners knowing about it. The next step could be that these computers would make autonomous purchasing decisions for evolving communication services: for example, "quality-of-service" (high bandwidth and throughput, low jitter and delay), "least-cost routing," "unpredictable results from peer computers." The prerequisite would be that they (yes, the computers!) have some form of revenue — for example, from humans wishing to pay for better person-to-person services. Computers could then become targets for some form of "advertising" by telecom service providers!

One can imagine a business model in which telecom service providers would embrace and develop peer-to-peer technology rather than combat it — for instance, peer-to-peer networks as the wholesale customers of the Next Generation Networks.

Not so unrealistic, since Skype, in 2005, selected four established international carriers worldwide, to terminate the SkypeOut calls to plain old telephone networks worldwide.

A next release of Skype and other peer-to-peer clients might automatically establish call segments (probably still packetized voice) over managed quality networks, in case low voice quality is detected. Skype might submit these calls to gateways, whose performances (traffic handled and voice quality increase) would be assessed and rewarded (e.g., in the form of SkypeOut credit). Voice quality could be measured technically (e.g., in terms of packet loss, latency, jitter) but also by a network of human test users granting Mean Opinion Scores (MOS) in real time.

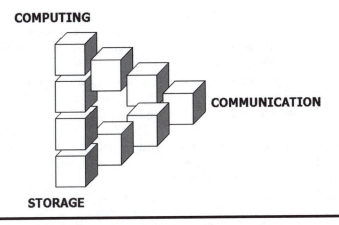

COMPUTING

COMMUNICATION

STORAGE

Figure 1.3 Storage is the cornerstone of computing, which is likely to fuel communication. But storage also pushes back the need for communication.

Managed communication networks and computing end-points may one day become better allies than communication and storage.

1.4   Means and Appropriateness

The explosion of number of available communication and storage methods requires the originator of the communication, but also the author of stored media to make frequent choices, based on the context of the communication.

The degree to which timely arrival, a response or unpredictable reactions (i.e., genuine communication) are desired by the originator/author will drive him/her to one of the readily available communication means or stored media transmission means.

The ultimate selection of some means requires judgment and careful decision. The receiver of the genuine communication or the transmitted stored media will indeed make a judgment whether the means are appropriate for the subject of the communication or message.

The appropriateness may simply be a matter of common sense, personal interpretation and company policy. "Do not send large attachments to our entire staff, but post your document on our intranet and distribute the link only," your IT administrator might say. "And use YouTube to post your funny joke videos rather than sending them around as e-mails to your colleagues."

In some cases, the choice could be so complex that it requires public debate, laws and/or court decisions.

Is it legal to fire an employee by sms? Has the sms delivery notification message (the proof that the sms was delivered to the mobile phone) got the legal value of an assigned letter in such case?

In the category of stored media transmission, is it appropriate to post someone's last words as an MP3 track on the Internet? Is it socially acceptable to set up a Web site regarding someone's death and funeral? Or is only the newspaper's necrology page regarded as socially acceptable?

How is law going to deal with the subtle distinction between distributing copyrighted electronic content by e-mail (legal), making it available for download on a home page (illegal), and all forms in between (e.g., peer-to-peer networks, blogs, chat sessions)?

Is it legal for a content provider to encode/protect the contents so it can be decoded/read only on a given computing device? Or be distributed only through a given telecommunications network?

At present, I am not aware of any law school teaching Digital Rights Management (DRM).

Is it a good idea to set up a Web site about your community, advocating the merits of its shops, restaurants and leisure clubs in good faith? If your site attracts many visitors, the baker might start giving you your Sunday sandwiches for free, for his shop to appear on top of the list. Is that socially appropriate behavior, or a new kind of corruption?

As the number of social conventions and rules also keep growing, the originators and authors could become tired of having to make such decisions of appropriateness. The combinatory explosion of mappings between communication means and social conventions is difficult to learn and to remember.

If the receiver's availability and preferences are ignored, it may also generate a sentiment of intrusiveness on the receiver's side and we'll discuss that in Section 1.8. But the point here is that the originator is tired of having to guess what communication means or content format will most please the receiver(s), be socially acceptable, legal, and so forth.

The result may be no genuine communication and no production of stored media at all: a post-Communication society comparable to today's post-Industrial Age.

Therefore, it may be time to ease the life of originators and authors, by shifting some of the burden and responsibility to the receiving end.

But have they (the receivers) got the choice today to select the communication means that are going to be used?

In genuine telecommunications, to a certain degree, the answer is yes. One can switch off a mobile phone (a radical method) and expect the callers to try other means (leave a voicemail, call a landline number or interact via instant messaging instead). Often, the callers won't. In an average fixed or mobile telecom network today, less than half of the calls are successfully established between two human end users.

For the transmission of stored media, the receivers have the choice of the delivery channel. For blockbusters, one has the choice to go and see the movie, rent the DVD, use digital TV (pay-per-view), or download the movie to a computer or set-top box (in divX format), burn it on DVD, and so on. But in most cases, the author will not be able to reproduce the content itself in many formats (e.g., as movie, book, comic strip, Web site, piece of art).

For genuine telecommunication services to be more successful, things could and should be improved on the receiving end. Receivers of genuine communications deserve a similar degree of choice as the receivers of stored media.

Telecom service providers would then take on the duties to let the user subscribe to, select, filter, translate, adapt, store, organize and reroute *incoming* communications and information, rather than today passively conveying the *outgoing* communication or information to the selected destination — a commodity.

This paradigm shift is probably too large a step for today's communication service providers to take, meaning that new, dedicated organizations will emerge which will successfully deal with these aspects.

The knowledge economy needs to tackle the problem of communication means and appropriateness, if it does not want to create a class of the super-communicating, super-informed, who can assimilate the information streams coming in through dozens of channels, and a class of isolated and noninformed people who feel flooded and anxious. We need a middle class too!

1.5 Contextual and Interactive Portions

The potential intrusiveness of incoming communications was recognized in the corporate and business environment, where human or automated telephone attendants, receptionists and call centers were introduced to handle the first part of the call, before connecting the caller through to the final destination (the correct department or most knowledgeable professional).

It is similar to the physical lay-out of many companies, with a reception desk where the employees provide the first layer of interaction with the external world. The first (and sometimes the only) impression that a company makes, is made at that reception desk.

Modern directory inquiry services are partially human, partially automated. After a welcome prompt, the system selects an available human attendant. He/she asks for the name or address, looks up the directory number, and immediately transfers control to the system, which reads the number by text-to-speech. The system then proposes to connect the caller through to the final destination, to send the number by SMS, and so on. It's been a cost-saving trend: a minimal interactive human portion is complemented with a growing automated portion.

Calling card services require you to call a free number, key in your card number, followed by the number you want to reach.

The most basic fixed telephone call today still starts with a sequence where the caller gets dial tone from the first (local) telephone exchange, dials the number and hears the ring tone produced by the last telephone exchange.

In the beginning of the twentieth century, before the advent of automatic telephone exchanges, the caller would be connected to a switch board attendant and ask to be connected to a given number.

Part of a telephone communication is contextual; part of it is truly interactive.

It is the mix of both which seems to make communication means attractive and successful.

When the communication is mainly interactive, the contextual portion on the other hand gives the originator the time to structure his/her thoughts. Didn't you ever feel confused when (by some technical fault) no ring tones were played but the called party suddenly took the call?

Some modern communication means that used to be purely interactive are evolving to also include a contextual portion.

Chatting and instant messaging, for example, used to be purely interactive. Some IM users then started to modify their display name every day, telling about their mood or upcoming activities. Later it was added in as a feature, separate from the nickname. IM/VoIP clients allow users to add a picture, some profile information, and so forth, that can be viewed by the caller before the actual communication takes place. Some mobile operators allow their users to select the ring tone that callers will hear (known as "Personalized Ring Back Tone" or "Colored Ring Tone") and this service is a smash hit in some countries (Korea).

Conversely, communication means that used to have only the contextual portion, are now adding the interactive part.

Most company Web sites (and intranets) used to be contextual only. But airline, insurance company and utility Web sites have been transformed to include an interactive portion, where people can fill in their phone number and click "Call Me" buttons. "Skype Me" buttons, which you can insert, for example, in your e-mail signature, make the persons who click this button use their Skype VoIP client to contact you. On some public and intranet sites, it's already possible to directly click other employees, departments or knowledge areas to set up voice-on-Web calls and video-on-Web sessions.

In the television space, there's a similar trend where viewers are invited to use SMS, e-mail or the show's Web site to participate to the quiz or give their opinion.

Of course people will use different communication means in parallel, in some particular situations. The typical example is chatting during a conference call or Webinar. Another one is photographing or filming the surroundings with your cam phone, and transmitting that as a video call, while continuing the voice conversation using your Bluetooth headset. These are referred to as *combinatory services*.

In the long run, the contextual part of the communication may thus become extended to cover the entire duration of the interactive part, as illustrated on the diagram below.

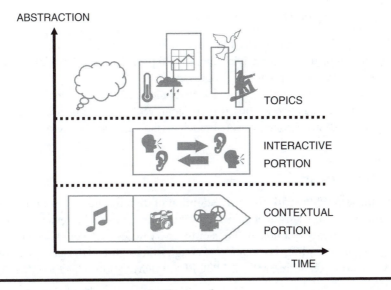

Figure 1.4   Structure of a communication session.

For a single, mainstream communication means to be successful, it should today probably always be conceived as a contextual part followed by an interactive part.

If not, the missing portion will have to be added as a bolt-on feature and this can never be as successful as incorporating it in the original design.

1.6   Device, Connectivity, and Service

Both true communication and stored media transfer involve consumers purchasing more or less tightly packaged offers including a device (e.g., a pocket PC), some connectivity (e.g., WiFi and DSL for Internet access) and some services: e-mail, Web browsing, instant messaging, search engines, VoIP just to name the most popular on the Internet. These three components can, together, be considered to form a "communication means."

As we will see in Chapter 6, developing and delivering an actual service to the end user involves much more than writing the software application and installing it on some server. Without entering into details here, a "service" can best be considered as an "application" surrounded by "enablers."

Is the consumer really in the driving seat to select the three components of a communication means?

1.6.1   Stage 1

In the 1980s and early 1990s, most decisions would have been made by the — then tightly coupled — connectivity and service provider, by the equipment manufacturer or by some corporate IT department.

The engineers in there would have made decisions on signaling standards in access and core networks, but also on "approved" end user devices (e.g., fixed analogue and digital phones, private branch exchanges, telematic terminals).

1.6.2   Stage 2

In the late 1990s, with the mass adoption of digital cellular communications, the communication consumer gained power through the mobile device manufacturers: no new network technology (i.e., connectivity or service) would become successful if it wasn't spreading down quickly enough from high-end devices to the lower end. A few examples were SMS, WAP, GPRS, CDMA2000, UMTS, CMDA1X, and MMS.

The joint selling and cross-subsidies between connectivity and device lead to strange technical solutions such as CDMA/TDMA phones without SIM, locking in the consumer to a given CDMA/TDMA connectivity provider. On GSM side, phones were programmed to lock in the SIM, forcing to use the connectivity provider which had subsidized the phone.

The joint selling of connectivity and service was also a common practice, resulting in 50/50 deals for premium SMS services, "walled gardens" for mobile data and a flourishing ring tone market. Service A could exclusively be used through connectivity A, and service B only through connectivity B.

This second period ended in a fragmented market of dozens of "communication silos" (tightly coupled device/connectivity/service combinations), with no new communication means *really* emerging — most people continued to only talk and text. Seen from a distance, the chance that mobile service providers would have succeeded to combine the *killer* device (O2s XDA?) with the *killer* connectivity (UMTS with roaming?) and the *killer* service (a music store?) was thin.

1.6.3   Stage 3

With the advent of Internet connectivity and more open operating systems in the early 2000s, even if the actual use of IP services on mobile devices is still in its infancy in 2006, a third period has started, in which the markets for devices, connectivity and services started to diverge.

The mobile device consumer's attention shifted from communication technologies to "local" functions such as cameras, storage cards and local connectivity (Bluetooth, WiFi). According to the Mobile Life Report 2006 by The Carphone

Warehouse, price comes second. Third are style factors: form factor (e.g., clam shell, slidephone, PDA), weight, design, and brand image.

The consumer of mobile connectivity lost attention for the complex tariff plans of mobile service providers, but was attracted by practical services such as new prepaid top-up methods, Web-based prepaid (offered by MVNOs — Mobile Virtual Network Operators), or single-stop shopping for communications, Internet, and TV.

The "Full MVNO" trend, where independent organizations operate their own HLRs and prepaid platforms, is a perfect illustration of how services can be split off from connectivity. Of course both layers continue to communicate in real time to control the routing and duration of calls, sessions, messages, and hopefully one day which media are allowed and what QoS is applied (we'll discuss that further in Section 3.3.2).

In such an environment, the choice of services can and will thus still be made later. Televoting, gaming, navigation, music download, community services and telemetry will not necessarily come from your current mobile and Internet connectivity providers.

1.6.4 Evolution

The cellular communications industry isn't quite there yet, but a similar evolution took place in the personal computing and Internet space, around 15 years earlier.

Since the early 1990s, the markets for computing devices, Internet connectivity and services evolve completely independently.

Specialized form factors (e.g., laptop, pocket PC, cube, server blade, set-top box) and designs appeared on the device market.

Connectivity technologies evolved from dial-up to digital subscriber lines, cable, wireless and soon mobile.

The services market is probably also 10 years ahead of what we see in the cellular space, where the usage of search engines for example, is marginal. The usage of Web-based address books is nil from cellular devices (people use the local address book in the SIM and/or the phone). But it is substantial in the PC/Internet world — for example, with free services such as Plaxo and LinkedIn.

New communication means should be developed and positioned either as a device feature, or as a connectivity mode, or as a service. It won't be possible to bundle the three aspects anymore.

1.7 Cellular, Mobile, and Wireless

In an ideal world, the service plane would be completely decoupled from the connectivity plane; therefore we wouldn't need to consider these connectivity properties when discussing services.

But the possible usage patterns of services are today strongly influenced by the degree to which this connectivity is cellular, mobile or wireless.

A good example is the current situation within IEEE* LAN/MAN Standards Committee (LMSC), where three competing long-range broadband wireless access systems have been standardized: 802.16-2004 (Wireless MAN, "WiMAX"), 802.16e-2005 ("Mobile WiMAX," February 2006) and 802.20 ("Mobile-Fi"). These three connectivity technologies can be differentiated by their degree of mobility (handover speed) and degree of being wireless (the range), among other important factors such as throughput per cell and spectral efficiency (i.e., bit/s/Hz).

By the way, it must be said that in the most recent simulations,[1] Mobile WiMAX, based on an OFDM air interface, achieves remarkable results in comparison to CDMA-based 3GPP HSPA and 3GPP2 EVDO. For the first time, for example, Mobile WiMAX will break the magic line of 1 bit/s per Hz and achieve downlink throughputs of 6 to 14 Mbit/s per cell (compared to 4 Mbit/s for HSPA and EVDO).

Of course, some nontechnical considerations will come into play as well (economics, politics). Mobile-Fi is technically better suitable for mobile VoIP (faster handover, lower latency), but more 3G-threatening due to its low frequency bands below 3.1GHz. And anyway, the actual implementation of Mobile-Fi would be further ahead, as Mobile WiMAX development is now almost completed. Therefore, the IEEE Standards Authority just decided to freeze the Mobile-Fi work for a few months.

1.7.1   Cellular

In a *cellular* network, small groups of devices share a single local resource (antenna, base station) based on their current geographic location.

IEEE Mobile WiMAX, 3GPP, and 3GPP2 are of course standardized cellular networks.

But cellular networks, in our opinion, are not necessarily wireless: a LAN can perfectly be considered to be a cell, with the IEEE 802.3 hub as the shared resource.

Even a Virtual LAN (VLAN), spanning multiple physical locations and using the IEEE 802.1Q-1998 tagging protocol, but no IP routing, can be considered to be a cell.

Examples of noncellular networks are DSL and cable networks, in which a dedicated resource (copper wire, coax cable) is used to connect the end device (modem) to the core network (DSLAM, CMTS).

1.7.2   Mobile

In a *mobile* network, devices/users have the rights to roam across multiple geographic locations within a same city, same country, across countries and — increasingly — across different networks, technologies and devices. Coming back to the example

---

* Institute of Electrical and Electronics Engineers.

of the enterprise network, such network will often offer a degree of mobility (I can plug in my laptop in any meeting room).

Handover is the process of keeping communications active across the boundaries of cells, regions and technologies. Handover needs to happen fast for a technology really to be considered mobile; for example, Mobile WiMAX will allow handover at vehicle speeds up to 120 km/h and Mobile-Fi up to 250 km/h.

If a technology uses more widely available frequency bands, such as Mobile-Fi compared to Mobile WiMAX, it can also be considered to be more mobile, as it has the potential to be deployed in more countries. Mobile WiMAX has been defined to work in five distinct frequency bands ("Release-1 System Profiles") between 2.3 and 3.8 GHz.

Let's note that some cellular networks are nonmobile: CDMA cellular technology is currently used in India to connect Public Central Offices (PCO, fixed phone shops) in rural areas. But specific software in the network will prevent such PCO devices to be moved to the next village.

Finally, some mobile networks are noncellular: digital video broadcasting by satellite (DVB-S and DVB-S2), for example.

1.7.3 Wireless

In a *wireless* network, the physical communication takes place over the ether.

Most networks today are wired. Even in a modern cellular mobile network, a call or message uses wires for more than 99% of the distance.

The general technology trend is even to lower the range which the wireless devices have to bridge (the wireless reach). Cells can then accommodate more users or deliver more throughput per user, fewer transmission errors, and less power consumption.

Health considerations and the use of unlicensed frequency bands will also force us to reduce power and wireless reach.

In this sense, the degree to which networks can be considered to be "wireless" is decreasing.

In developing countries, the demand for wireless access will be stronger due to the relative scarcity of copper wires, coax cables and fiber glass.

1.7.4 Trends

More than the need to remove a popular confusion between these three aspects of modern communication, it is important to be able to position communication technologies along these three axes and to notice trends.

Technologies which used to cover the three aspects (e.g., UMTS) may be attacked by cocktails of more "specialized" technologies (e.g., DSL + WiFi + Mobile IP). DSL technology does not need to consider cellular aspects, WiFi standards are not polluted by mobility, and Mobile IP does not worry about cellular or wireless.

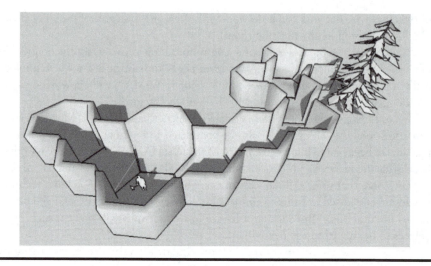

Figure 1.5   Cellular mobile wireless networks can fence you in!

Typical mobile procedures such as in-call handover may leave the realm of standardized technology. To perform good in-call WiFi-to-GSM handover may not require a new 3GPP standard, but may require fine real-time analysis of the voice quality, for example.

Is there a trend toward more cellular technologies such as HSPA, against noncellular such as VDSL? No. IP rules the world, and uses whatever underlying technology to get to the end user.

Is there a need for more mobility? Or have GSM and CDMA fully satisfied our demand? Well, new demand for mobility could for example come from in-vehicle devices such as car entertainment systems, the on-board computer or the alarm system.

Are we going more wireless? For sure, there's a demand for broadband wireless Internet access in emerging markets, but also covering cities, trains, petrol stations and holiday resorts.

1.8   Presence and Availability

Timeliness, expectation and unpredictability are the cornerstones of genuine communications.

However, these three aspects can create an "on call" atmosphere in which communication users feel forced to be reachable within the shortest possible time at unpredictable moments and in inconvenient situations.

As a reaction, they might start shutting down their mobile phones, mobile e-mail clients and computers for too long, and might (probably after a few bad experiences) develop a sentiment of anxiety to miss an important communication or piece of information — either good or bad news.

In the end, the user may feel unsatisfied both when the communications device is on and off — something all telecommunications and information providers should of course try and avoid at all cost.

To counter-balance the sentiment of intrusiveness, the root cause of the trouble, modern communication means should be able to track and take into account the presence, availability and preference of the user on the receiving end.

Presence was largely ignored by the Plain Old Telephone System (POTS); it was impossible for the receiving user to publish his/her presence, namely to signal that he/she was or wasn't physically in the neighborhood of the black phone. Hence the caller simply had to try and wait for someone to pick up the horn.

Answering machines and Interactive Voice Response (IVR) systems started taking calls in the 1970s, and people would have them switched on even when they would be around — either at home or in the enterprise. This may be interpreted as an expression of protective attitude against intrusiveness (you could delete these unwanted messages, filter out callers not having a customer number, etc.), but also the attractiveness of deferred communication (the messages could be answered later). We'll address immediate vs. deferred communication in Section 1.9.

Basic presence is sufficiently addressed by existing mobile cellular networks, and this convenience has most certainly contributed to their success. Mobile phones may be switched off, set to vibrate and to silent mode. The mobile network can forward incoming calls in case the called person is unreachable, busy, or doesn't answer the call. The phone's on/off status is known by the network, but once the handset is powered on, the decision to answer the call can still be made based by the called user (based on who's calling: Calling Line ID Presentation).

Fixed enterprise telephony solutions (Private Branch eXchanges, but increasingly Centrex and VoIP Centrex) offer advanced incoming call handling. Automated or human attendants can limit the intrusiveness. Increasingly, these attendants are able to take into account of calendar information (appointments, meetings, holidays). Call centers are the school example of the practical use of presence and availability information to handle inbound calls.

Company e-mail servers allow you can set an automatic reply message, when on holidays. Web-based e-mail servers usually don't. On the other hand, as with your letter box, your e-mail server can't refuse to take an incoming e-mail. Therefore, lists of senders and keywords can be set up to filter junk e-mail. The IETF developed an e-mail filtering language called Sieve (RFC 3028).

Instant Messaging programs offer a good view on the presence and availability of buddies in a buddy list, but then again not on their presence on other devices (mobile phone, gaming console, IPTV set-top box).

The availability of the user is sometimes discovered by the device itself — for example, the availability is set to "Away" when the keyboard and mouse are not touched for a while.

Some IM programs now allow logging in as invisible, which shows that people are maybe not that willing to show their actual presence anymore. Could the solution be to publish different presence and availability to different groups of buddies?

Other communication means have until now remained presence-, availability-and preference-agnostic: SMS, for example, does not allow the sender to see the destination's presence in advance of sending the message.

On the inbound side, SMS does not allow you to set an autoreply, to forward inbound SMSs to another terminal, or to bar messages from some sources.

Are presence and availability going to spice up existing communication means? They could, for example in the case of SMS, or even the POTS.

If your PC detects keystrokes or mouse movements, this could be communicated to a presence server in the POTS, assuming that your telephone is on your desk next to your PC. Callers could then hear that you are away, if you didn't use your PC for say 10 minutes.

Conversely, if the upgraded, presence-aware fixed telephony network detects that you make or receive a call, it might be configured to inform your IM provider, so your IM contacts can now see that you are "at work."

Or are presence and availability going to enable completely new communication means?

For sure, presence and availability information would pave the way for "push" services. Today, most content and services are consulted in "pull" mode, in which the end user has to take the initiative. If the user could let his/her presence and availability be notified to a selected set of applications, these applications could send "fresh," relevant content, rather than continuously pushing down needless streams of information that are going to remain unread in some Inbox. This notify-and-push principle exists in an embryonic format on the Internet, as "syndication" (even though today the RSS Web feeds — Really Simple Syndication — are in fact a pull technology).

Selectiveness and trust are essential; users would be reluctant to let their presence information be notified to all contacts, tribes and applications, for unknown (worse: undesired) communications and content to be pushed down. Similarly, today people are (rightfully!) increasingly reluctant to provide their contact details (e-mail addresses and telephone numbers).

Obviously this sector will mature, and the presence and availability of information may be merged into a larger functional entity such as personal homepages, Web-based address books or social networks.

Communication service providers should consider presence and availability as one of the healthy ingredients in the cooking recipe for any new communication service.

They should inspire the confidence that the users will always be able to do the following:

> Grant and revoke permissions to their personal contacts, tribes, and applications to watch their own presence

Make sure that their own presence data will never be watched nor modified
by nonauthorized individuals and applications

Handle incoming communication based on their presence and availability data

1.9   Immediate and Deferred Communications

Maybe as a consequence of the protective measures against intrusiveness, communication users have discovered the benefits of deferred communications.

We saw that timeliness is essential for genuine communication to take place: a short time span between the production and the reception of information.

When it used to be humans receiving the information, machines (e.g., answering machines, IVR systems, voicemail systems, e-mail clients, SMS/MMS-enabled mobile phones) are now receiving these calls and messages. Humans now have the comfort to read, interpret and react on them later.

Communication is deferred, and sometimes buried among stored information requiring no action.

Has it therefore become less "genuine"?

To some extent, yes. In large companies, there's today a noticeable erosion of the value of an e-mail, for example. Senior executives and their assistants receive, skip, archive, interpret and send hundreds of e-mails a day. A certain error rate is accepted by the originators. The expected time of reaction is, after having gone down since the adoption of e-mail and mobility, increasing again. Senders "escalate" by using multiple communication means, which is worsening the problems. How many voicemails start with "Hey I sent you an e-mail," and vice versa?

Are users expecting their communication service providers to play a role in improving things, by sending reminders for pending messages or actions? Perhaps, if these notifications could be limited and aggregated (providing a daily summary instead of a continuous stream).

Perhaps someone could also suggest priorities to all these messages, by learning from the priorities we assigned ourselves, our past reaction time to that sender, and so on.

As for most basic forces structuring modern communications, value-added services relying on them will become more successful if they can be provided spanning multiple communication means. Today this is not the case; you would be notified by your cell phone, your instant messenger, your online game, and so forth.

The idea of the communication service provider as a personal Inbound Communications Handler (ICH) will come back throughout this book.

In the meanwhile, our pending and archived communications are stored in giant data centers around the world, for example, in the order of 3 PB* for

---

* A petabyte = more than 10^{15} bytes = more than a million billion bytes.

a 50,000-person company, or 60 GB* per employee. Across all communication means, world demand could be estimated to be in the order of 5 EB† today.‡

Further demand could arise from storing media in these data centers; that is, if people would be allowed to leave all their e-mails and attachments on the servers, or if video would start prevailing over voice communication, we would be on a steep curve to the ZB.§

Whereas this calls for advances in database technology and economies of scale, a second aspect are the networks to access all these messages. Broadband connectivity providers are perhaps not connecting users to other users, but users to their message library.

The evolution to deferred communications could end up in a world where messages and multimedia content are stored and universally accessible, and people communicate in terms of "links" (pointers) to stored content. A world where knowing how and where to find the info is a more precious skill than knowing much or producing much new information.

In that sense it would be useful for people to be able to talk about these pointers; for communication service providers to invent and promote extensions to language. Universal Resource Locators (URL) are obviously not appropriate.

After all, haven't SMS and IM users invented their own "pointers" to longer expressions?

1.10 On-Demand, Triggered, and Periodic

The circumstances in which communications are launched (originated) have an increasing influence on their frequency.

It's a pretty vague statement, so let's pick an example.

Newspapers, for example, are available in three modes: on demand at a newspaper outlet, triggered by events such as major catastrophes, or periodically, in subscription mode.

On-demand communications require repeated purchase decisions by the end user, therefore limiting commercial success in the long term, certainly in a world where communication means are abundant and compete with each other. Their benefit, on the other hand, is that they look less obtrusive.

---

* A gigabyte = more than 10^9 bytes = more than a billion bytes.

† An exabyte = more than 10^{18} bytes = more than a billion billion bytes.

‡ According to a September 2006 report of industry analyst IDC, shipped storage capacity grew to 704 PB in the second quarter of 2006, representing 51.5% year-on-year growth. With an average system lifetime of 12 quarters, total installed capacity should be at around 7 times that number (4 + 2 + 1).

§ A zettabyte = more than 10^{21} bytes.

Triggered communications are a relative novelty. SMS, e-mail, IM and RSS are examples where the communication is accomplished when the recipient user "surfaces" (powers on the device or launches a program). Different events could be imagined to trigger such communications, from world news, over stock quote changes, a car accident, to a particular location or mood of the recipient user. The possibilities of such communication means are yet to be explored fully. What is their degree of perceived obtrusiveness in the long term?

Periodic communications are probably the most obtrusive, if established between humans. Even if touted as more acceptable between machines (e.g., for telemetry), cost-conscious communication consumers (i.e., of mobile data megabits) are looking for less "chatty" applications. Consumers may end up feeling trapped in a system they subscribed to, with no possibility to cancel. The periodic mode is probably more suitable for the transfer of stored information, than for true communication services. Online gaming providers who impose to their end users to be online at certain times of the week could soon discover the negative effects of periodicity.

When conceiving new communication means, the on-demand, triggered and periodic aspects need to be evaluated in great detail.

The role of the communication service providers could be expanded from granting permissions for on-demand services today (calls being made thanks to an existing billing relationship or prepaid account) to storing all information regarding triggers and subscriptions to communication and information services, tomorrow.

It is just another weapon in the battle against the growing commoditization of connectivity, communication and information.

1.11   Contact Lists and Tribes

During the course of their social lives, individuals build personal contact lists which are today maintained on mobile phones, on PCs, and on the Internet.

According to the Mobile Life Report 2006 by The Carphone Warehouse, mobile phones have allowed us "to return to the more natural and humane communication patterns of preindustrial society, when we lived in small, stable communities and enjoyed frequent communication with a tightly integrated social network of family, neighbours and friends."

The report further finds six socio-demographic segments which it labels as "Tribes" (Generation Mobile, Phonatics, Practical Parents, Smart Connected, Finger and Thumbs and Silver Cynics). These are, however, merely market segments: though these people must certainly have a common usage pattern of mobile phones, they don't necessarily communicate more with people within the same segment. Also, it is difficult to say whether or not these segments are also applicable to different communication means such as instant messaging.

Through our social life we become members of schools, companies, sports clubs, interest groups, political parties and unions — "real" organizations. In the

Internet era, we subscribe to newsletters, use Web applications (banking), create Web passports and log in to various machines and applications, and become members of Web communities. The distinction between the former organizations and the new "electronic organizations" is getting more blurred every day. A generic name for these groups of users could be "tribes."

The ability to maintain these contacts and tribe memberships has become an important aspect of everyday life; who would like to lose his/her personal login/ passwords file without having a backup? Who has never experienced the frustration of transferring the contacts list from an old to a new mobile phone? Or to log in to your company's Web mail portal and discover that you can't send e-mails because your contacts are not in there?

1.11.1 Contact lists

The Internet Mail Consortium's standard vCard format is an XML document allowing storage of a personal contact's e-mail addresses, telephone numbers, home pages and geographic address. It allows the PC's address book to be synchronized with most mobile devices.

As for presence and availability, one can wonder whether this information belongs only within devices or also within the network.

Plaxo, for example, offers individuals the opportunity to copy personal contacts to a Web application, and to keep them synchronized.

Web-based e-mail programs rely entirely on Web-based contact lists.

People however are not ready for a Web-only contacts list: synchronization can be a tedious task, but it should remain possible to watch the contacts without having network access.

1.11.2 Tribe Memberships

How about the tribe memberships?

Let's here consider a tribe from the strict communications sense: the list of users of a given service.

Today the tribe's own member list is maintained in (hopefully a set of redundant) servers.

Tribe members themselves have their own schemes (from text files to spreadsheets) to maintain their personal list of memberships. There are no standards (yet?) to store and synchronize this information between the tribe and the member. Higher security services such as Web banking now even require additional physical signature devices.

Without local address books, people would make fewer outgoing calls and send fewer text messages from mobile phones.

Would it therefore be a fair explanation of today's low success of mobile data services, that there's simply no tribe membership file on today's mobile handsets?

That there's no "single sign-on" procedure to the services, except maybe to the services in the walled garden of the connectivity service provider?

If these contact lists and tribe memberships are well thought of, it should be possible to organize them in the following ways:

> With multiple identities: I may be identified by application X as a company, and by application Y as an individual.
>
> With multiple profiles: I may not wish to reveal my personal vCard in a professional context, and vice versa.
>
> By category: it is difficult on today's phones to organize contacts in various categories (e.g., family, friends, colleagues, recently used), and to keep that organization while synchronizing.
>
> With extensible attributes: it should be possible to add future contact details through new communication means (e.g., the SIP address).
>
> Relationally: it should be possible to jump from contact to contact, along relationships representing the real structure of our social networks (e.g., brother, employer, coach of).
>
> With two-way handshaking: when contact B is added in person A's address book, it should be possible to invite person B to add contact A in person B's own address book.

1.11.3 Evolution

As in the case of presence and availability, various industry players are going to fight to hold contact lists (historically the domain of mobile phone manufacturers and e-mail client producers) and tribe memberships (e.g., for the production and distribution of security certificates).

Maybe a new industry will emerge.

Alternatively, contact lists and tribes will migrate from the domain of e-mail and telecommunications into the realm of various industries related to authentication, security, identity management and electronic purses.

1.12 Location

Much has been said about location and its benefits, for example the possibility to adapt content, tariffs or incoming call handling based on the user's current location.

The mobile communications industry has tried to assess which granularity of location information would be required for which services. For sure, navigation requires the granularity of GPS. But could so-called home/office zone services (lower tariffs) be provided based on the GSM cell ID, in an attempt to displace the landlines?

Would people be willing to pay a premium to find the nearest pizza restaurant in Paris (an iconic example at GSM congress presentations), rather than just asking someone?

Some people started to think about location-based services (in mobile networks) as a remedy to poor user input interface on the terminal (small display): "the mobile device would automatically display the map of where I am." OK, but that's precisely what a GPS navigation system does.

Or as input data for push services: "walk into the shopping center and receive a targeted advertisement," another example in the same series.

A few niche services have been launched, but in general, location based services have not delivered on the promise.

Worse, the market is so immature that an Italian operator saw a sudden decrease of subscriber numbers once their current location became available on the Internet, even if previous permissions had to be granted by the mobile subscriber. People did not even want to enter the debate on whether they would grant the viewing permissions or not; they simply cancelled their subscription to get rid of the problem.

In the recipe of a new communication service, I would therefore qualify location as the garlic: excellent for health, deterring to many, and always to be used in extreme moderation.

Chapter 2

A SWOT Analysis for Communication Service Providers

In this chapter, let's land our chopper on the grounds of the Communication Service Providers (CSP): Fixed and Mobile Operators, including new entrants such as Quad Play* Service Providers, Instant Messaging and VoIP Service Providers, and Web-based Mobile Virtual Network Operators (MVNO).

Let's consider their assets in terms of strengths and weaknesses, and the opportunities and threats offered by the immediate surroundings of our landing spot.

2.1   Strengths

Today, Mobile and Fixed Operators have highly valuable assets.

GSM/CDMA cellular radio **coverage and copper** lines have both proven to be able to carry IP traffic in always-on mode, using GPRS, CDMA2000 and DSL technologies. The bandwidth is sufficient for many services, and can still be increased, at the cost of huge investments though (3G on the mobile side, VDSL on the fixed side). This should enable streaming services such as digital radio, digital TV and video calls to be carried over these packet networks — services which put quite stringent requirements on the underlying bit pipes in terms of Quality of Service:

---

* Simultaneously providing fixed, mobile telecom, Internet access and digital TV.

high availability, throughput, latency, jitter and error rate. This transport Quality-of-Service is difficult to offer on copper pairs you don't own, or using unlicensed frequencies.

In the fixed network, the unbundling of the local loop has allowed Internet Service Providers (ISPs) to offer broadband Internet access to the residential and business communities. But the last mile of copper has remained the property of the telephone company.

Second, Mobile and Fixed operators share the valuable asset of the **E.164 addressing** space, and, related to this, the existing commercial customer relationships for postpaid or prepaid telephony. New NGNSPs could be tempted to invent a new address space (personal nicknames, e-mail addresses, etc) which do not allow valuable inbound calls and messages (i.e., no inter-operator accounting). If they want to use E.164 numbers, these NGNSPs could have to crawl to painful issues such as licenses, interconnections and number portability — the key skills of Mobile and Fixed operators.

Mobile and Fixed Operators have proven to be able to market messaging services on a **per-message** basis (SMS and MMS; person-to-person, person-to-application and application-to-person) — as opposed to ISPs and new NGNSPs, which have to let instant messaging services be used for free (at least to the end user).

Last but not least, Mobile and Fixed Operators have gained great **in-house experience** with the core network aspects of IP technology (e.g., addressing, routing, QoS, security) and packetization in general (e.g., WAP, GPRS, MMS) — maybe as opposed to ISPs, who have concentrated on providing higher-level services on top of basic Internet access (e.g., Web design ad hosting, e-mail, DNS).

2.2 Weaknesses

Most Mobile and Fixed Operators face some internal structural weaknesses, when attacking the market of new NGN services.

Two **rating and account management** systems are still in place: typically an online IN-based system for prepaid subscribers and an off-line billing system for postpaid subscribers, inter-operator accounting, and many more functions. Few Operators have gone through the organizational and technical process to rationalize this situation.

At Mobile Operators, Value-Added Services (e.g., voicemail, SMS, WAP, MMS, Prepaid, VPN, PoC) have been deployed as vertically integrated **stove-pipe applications**, each with their own network interfaces, databases, provisioning, billing, supervision and CRM mechanisms. The intricate Web of cross-vendor interconnections is now becoming an impediment to launch any new application in such an environment.

Whereas many Mobile Operators now belong to supranational groups, important processes such as purchasing, engineering and operations have shown to be too difficult to centralize (or to distribute to specialized centers). NGN technology clearly has the potential to be deployed at supranational level, but the **lack of technical group strategy** could hamper such deployments.

Finally, the **lack of collaboration between Mobile and Fixed** operators is a handicap, when facing NGN service providers, for which the distinction between mobile and fixed are just access methods (i.e., a different version of their SIP user agent software). Fixed-Mobile Convergence (FMC) is the key objective of the IP Multimedia Subsystem, but will the Operators be able to introduce the corresponding organizational changes?

2.3   Opportunities

This being said, NGN technology will open a series of opportunities for Mobile and Fixed Operators.

Quad play (the joint sale of fixed and mobile telephony, Internet access and broadcasting) should allow increasing revenues, even though NGN technology is not a prerequisite to realize it. Introducing NGN technology is just an additional opportunity to streamline the offer in these three domains (e.g., to offer one prepaid account or one bill).

Virtual Operators (low-cost, Web-based, prepaid, targeted at a niche segment) is a new way to market telecom services under various brands. Any new NGN service platform should be conceived to be shared by multiple service providers.

The great commercial uptake and natural attraction of residential broadband Internet access is a great opportunity for Fixed Operators in general. Adding a NGN and VAS layer on top should enable them to **move further up the value chain**. For example, instant messaging users are accustomed to the convenience of availability (e.g., Away, Busy, Appear Offline). In the next future, they will demand subscription-based and push-type telecom services that take into account their willingness, activities, preferred communication types and devices, calling buddies, location, agenda, time zone, mood, and so on. This personal policy should be stored in an open environment shared by multiple services, companies and even industries. Personal contact pages will entertain and inform the caller with personalized ring-back tones, pictures and animations, selected by the called party.

NGN technology should also allow the Operators to **reverse the business model** for some services. We will look at this in further detail in this book (i.e., in Sections 4.8 and 5.6), but the general idea is that not all revenue should come from the end user of the VAS.

Further specialization by layer (e.g., into Access, NGN Core, VAS and IT cost centers) is an opportunity for Mobile and Fixed operators to fundamentally **reduce the cost to serve**. A commercial alliance of WiMAX access provider, NGN core service provider and MVNO makes perfect sense, but each of them needs to focus on their core competencies and be profitable at a larger scale. A single company will fail.

These NGN, VAS and IT service providers may choose to **ensure service continuity** across legacy (POTS, GSM) and new (NGN, IMS) Access networks. An example of this is presence-based routing of inbound communications, which could also take into account the presence events on POTS side — for example, an incoming telephone call currently being answered.

Finally, it is generally expected that NGN/VoIP technology will be the trigger for corporate organizations to **outsource** their internal communication services to networked PBX ("IP Centrex") operators, which the fixed line operators (with experience in Centrex) are definitely well placed to become.

2.4 Threats

High-quality **peer-to-peer** services (e.g., for instant messaging, address books, file sharing, telephony, video telephony) today threaten the concept itself of a Network Operator. Residential broadband access seems to have been Pandora's Box and the results are unpredictable for the telecom industry as a whole.

With the phenomenal uptake of broadband IP in the home,* the broadcasting industry (based on coaxial cable) and telecommunication providers (with DSL on copper pairs) are engaged in a fierce battle for market share. Seen from the other side, the entertainment industry (including TV, movies, music, sports, and so on) now has a new do-it-all channel to the end user, called IPTV†. But will entertainment mainly flow through **IPTV, or just over IP?** The variety of content and richness of user experience on the Internet is staggering, and youth stopped watching TV.

Internet communities have huge commercial power, as they are organized by groups of people sharing the same interests. Telco's do not know their customers so well (their knowledge stops at an E.164 number, a calling pattern and sometimes a billing address). Broadcasters don't know their customers at all (except a billing address). Advertising revenues will therefore flow more easily to a targeted audience within these communities‡, rather than down the SMS/MMS, NGN/IMS or IPTV pipe.

But also in their access networks (the last mile), Mobile and Fixed Operators will face competition from new entrants using new wireless broadband access technologies (WiMAX and OFDM-based in general) over **unlicensed spectrum**. Please note that the WiMAX specifications support both unlicensed

---

* Over 70% of American homes will have broadband connectivity by 2009.
† Though this channel is mainly offered by telecom service providers (the Incumbent and Competing Local Exchange Carriers) today, it is clearly not a communications service but the transmission of stored content (cf. Chapter 1).
‡ People searching for "Gevrey-Chambertin" can automatically be considered to be member of the Bourgogne wine amateurs' community, therefore the search engine will make sure that the right sponsored links appear in the search results, and even on the target Web site.

spectrum (in the 5–6 GHz range) and licensed spectrum (below 11 GHz or in the 10–66 GHz band) — we'll review the standards in Section 3.7. Short-range unlicensed spectrum technologies (e.g., WiMAX, WiFi, Bluetooth, Wibree) may cause a move from Mobile (2.5G, 3G) to Fixed access (DSL, cable).

Finally, the third billion mobile users, to be connected before 2010, will yield much **lower ARPU** (average revenue per user) than the first and second one. If the Operators do not succeed to reduce costs structurally (i.e., with all-IP networks and converged billing systems), going after this next billion users could turn out to be nonprofitable.

2.5   Conclusion

Introducing SIP/NGN technology will allow Mobile and Fixed Operators to mitigate their Weaknesses and fully exploit their Strengths. But it won't influence the important new Opportunities or Threats that they are currently facing: these commercial battles have to be fought.

Multiple industries (e.g., telecom, broadcasting, instant messengers, search engines) are currently converging to the use of these consistent technologies.

Standardization is a way to create global, successful markets for these industries.

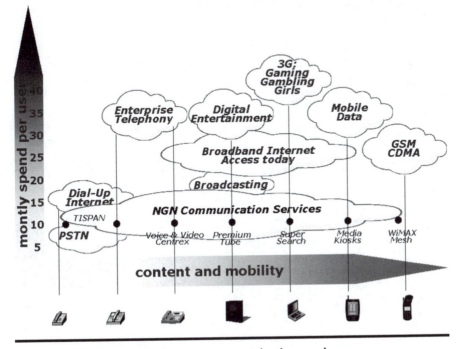

Figure 2.1   Opportunities for NGN communication services.

It creates the landscape, the battlefield in which the competitors are going to differentiate themselves with service offers.

The following diagram presents the commercial landscape for connectivity, communication and content services, in a typical, mature market.

Consumers and enterprises are willing to pay for content and mobility, but historically, competing technologies and devices had divided the market in very distinct industries such as broadcasting and GSM.

The unified NGN technology will allow addressing new segments — literally by *undercutting* most of today's technologies and communication methods in price, but also by providing a new customer experience across the different terminals.

On the far left and the far right side of the diagram, communication consumers might be willing to give up some services, in return for the broad, horizontally unified user experience. On the left-hand side for example, the electric power on an analog telephone line, which guarantees emergency calling in all circumstances, might be sacrificed in the NGN. On the right-hand side, consumers might be willing to sacrifice the national coverage of mobile cellular networks, in return for higher wireless speed, content or even social networking at well-located digital media kiosks.

In the next chapter let's review the past and ongoing standardization work for this NGN. Despite our efforts to summarize, interpret and position these standards relative to each other, we have to warn that the next chapter remains the most arid part of this book, and contains the most acronyms.

Fasten seat belts!

Chapter 3

Review of the Standardization Work for Next Generation Networks

In this chapter we take a tour through the work of the standards bodies that are currently driving Next Generation Networks.

When a telecommunication technology is successfully standardized, it results in a world market for devices, services and business models.

At first sight, it may seem an arid world of acronyms, architectures, procedures and protocols, driven by strong industrial interests.

But people should try and form a personal opinion: not only engineers and managers at communication service providers and their equipment vendors but also the consumers of devices, connectivity and services — a large group.

Is it still worth reviewing technology standards from the perspective of a communication service provider only?

Yes, insofar as these technologies could lead to operational efficiencies, better business models and a better end user experience. Yes also if introducing these improves the SWOT-situation of the communication service providers, in general.

Should businesses buy a 3G phone fleet now, or wait for IEEE Mobile WiMAX to be built into the next laptops?

Are nonstandard IM/VoIP clients going to be marketed as freeware, adware, and spyware, versus standard SIP clients under a licensed, pay-per-call, flat-free model?

Should the car industry adopt communication standards such as SIP today, wait for new ones, or develop their own? Considering their recent (and maybe not so successful) experience in navigation systems ....

It helps to look at the new standards from the perspective of past success formulas. Are the features still there (e.g., echo cancellation for voice)? Are known issues and imitations of past technologies (e.g., long call setup delays) successfully addressed in the new one?

Some services such as video calling were even developed throughout multiple standards (ISDN; 3GPP Release 1999 circuit switched; SIP), which helps to form an opinion on the benefits of each standard.

Also, uncovering the main drivers behind an organization's work one can shed some light on the usefulness of a technology.

But before we dive into the organizations and standards in the area (in order of relevance), it is time for some first attempts to define a Next Generation Network.

Not too broad, not too narrow.

3.1 Definition of a Next Generation Network

A Next Generation Network is a set of servers that allows User Agents to establish and be charged for Sessions, within which they can convey Media streams.

This statement is too trivial; it's like defining the Internet as a network in which we use HTTP to browse sites hosted on Web servers — which ignores the myriad of other applications on IP.

Under this basic NGN definition, user agents may, once they have established the session, exchange media streams freely, outside of the control of the Next Generation Network Service Provider (NGNSP).

We will use this new acronym, NGNSP, throughout the book, in analogy with ISP (Internet service providers) mainly providing connectivity to the Internet (but also e-mail, hosting, name servers and other services) and CSPs (communication service providers) providing traditional fixed and mobile communication services. For the moment, let's park the question whether NGNSPs will ever be part of the CSP family.

One can wonder whether a NGNSP is required, and could claim revenue, for such simple task as setting up SIP sessions to telephone numbers at domains.

Some people have even started to think of SIP as a peer-to-peer protocol, removing the need for a NGNSP to switch (analyze and route) these addresses.

Would it help to add the following conditions to the definition?

The Media streams will be forced to use a Border Gateway of the same organization as the one charging for the Session establishment. This NGNSP will take the responsibility to deliver adequate Quality-of-Service, according to the codec negotiated for the media stream.

OK, this would justify the existence of a NGNSP, only *if* Quality-of-Service (QoS) turns out to be an issue in peer-to-peer mode and if that issue cannot be resolved by just installing more transmission bandwidth and more IP routing capacity. Peer-to-peer services are, by the way, very good at consuming a larger portion of the available bandwidth than client-server applications (or spoke wheel architectures).

At the heart of the discussion about whether we'll see more NGNSPs or more ISPs in the future lies the question of whether the NGN can eat away enough Internet bandwidth from peer-to-peer applications.

Today, an ISP installing additional bandwidth only improves the quality of peer-to-peer services — not necessarily that of other applications.

The abilities to make outgoing communications to any destination outside of the NGN, or to receive incoming sessions from the outside world, are also not justifying the existence of a NGNSP either. Today these services are already offered to SIP users directly, by outbound VoIP carriers and inbound TDM collectors. Sometimes these companies are not even establishing public interconnections or operating media gateways themselves anymore.

Another aspect of such definition is that it does not differentiate the NGN from the "Current Generation Networks" such as ISDN, the Integrated Services Digital Network from the late 1980s (Section 3.6.1), on which GSM networks are also based.

The ITU-T's ISDN could be used to set up various types of channels (speech, fax, and circuit-switched data such as video) which could be considered as media streams. Strong QoS was guaranteed by the fact that these media streams were conveyed on dedicated circuits — which is why NGNSPs would deploy QoS-enabled border gateways.

Differentiation of the NGN is possible only with an extra (architectural) condition in the definition:

In an NGN, there's an open interface between the Session Control Layer and the Media and Border Gateways.

But is the communication service consumer going to care?

And if the session control and media gateway planes are split in the core of a traditional GSM network — splitting up today's MSC in an MSC server and a media gateway — does it then become a Next Generation Network?

It could be worth examining whether different organizations could deploy the session control layer and the media and border gateways. Probably this would lead to economies of scale, but slow down the implementation of QoS enforcement.

In conclusion of this little definition exercise, please find below an overall diagram defining the architecture of the NGN, to our best current knowledge.

This diagram should allow us to position the work of multiple standards bodies which are active in the telecommunications field.

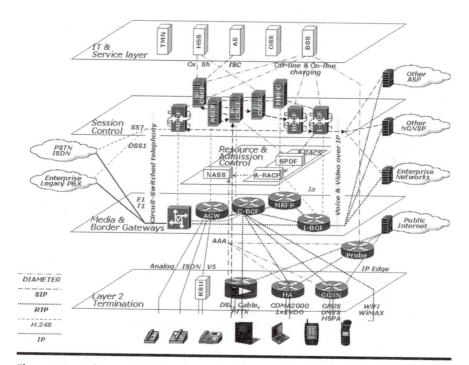

Figure 3.1 A layered definition of the Next Generation Network.

Under our definition, an NGN consists of five logical layers:

1. An IT and Services layer, containing the
   - Telecom Management Network (TMN) for supervision
   - Home Subscribe Server (HSS), supposed to become the master database and mobility manager for all subscribers in an IP Multimedia Subsystem (IMS)
   - Application Servers (AS), executing subscriber-dependent service logic
   - Operations Support System (OSS), the IT system for subscribers, product catalogs and provisioning processes
   - Billing Support System (BSS), the charging system receiving network information after the facts (off-line) or before the facts (online)
2. The Session Control layer, conveying the necessary signaling to let end users establish multimedia communication sessions
3. The Resource and Admission Control layer, allocating bandwidth resources to individual media streams, and admitting or refusing users to the NGN
4. The Media and Border Gateways layer, conveying the IP packets containing encoded media (voice, video), performing the necessary conversions and enforcing the policies decided by the Resource and Admission Control layer

5. The Layer 2 Termination, where devices such as a Broadband Remote Access Server are terminating layer 2 tunnels to the customer premises equipment (i.e., a DSL modem), but also of course using various wireless technologies

From a service provider perspective, this diagram is also designed to better illustrate "real life" rather than only their own network: interconnections to other service providers, to enterprises, northbound integration, the competition from non-SIP services on the public Internet are items that do not appear in traditional telco-space standardization diagrams.

Even though that won't be a major point of attention in this book, IP probes containing content filters, policy enforcers, and traffic metering functions will need to be integrated in the NGNSP's admission control and billing systems — certainly if the NGNSP organization (the communication service provider) is also playing the role of the ISP (Internet service provider). However, for sake of simplicity and in order not to blur our NGN definition with non-SIP-based services, let's not consider such equipment to be part of the NGN itself.

The interconnection to peer networks and applications deserves special attention, due to the security risks involved. Let's discuss that aspect later, in a dedicated Chapter 4.

Chapter 5 will address the charging and rating requirements for communication means in the NGN.

We will take a closer look at the Service layer in Chapter 6.

3.2   IETF

The Internet Engineering Task Force can be considered to be the founder of the Next Generation Network, but, as we will see, not the author.

Its Internet, Operations and Management, Routing, Security, and Transport Areas are laying the groundwork for a solid NGN.

More specifically, the Internet Area delivers a network layer, which can be used for both session signaling and media transport.

The Real-Time Applications and Infrastructure Area (a confusing name) standardizes the session and media layers.

The Applications Area deals with e-mail, Web authoring, calendar sharing, and so forth, but not specifically in relation to an underlying NGN. Therefore, we won't discuss its work in this book.

The IETF also created SIGTRAN (Signaling TRANsport), a set of important standards for the reliable transport of signaling over IP networks. The stream control transmission protocol (SCTP, RFC 2960, year 2000) laid the groundwork to carry signaling protocols requiring the reliability of the Transmission Control Protocol (TCP, RFC 793, 1981) without its slowness in networks with packet loss.

SCTP was designed to carry adaptation layers for ITU-T SS7 (see Section 3.6). But today, SCTP is considered as a viable alternative to UDP and TCP, in order to carry SIP (RFC 3261 in 2002 uses UDP; RFC 4168 in 2005 proposes SCTP) but also DIAMETER (RFC 3588 in 2003 used TCP).

3.2.1 Internet Area

The Internet area puts a lot of energy into mobility management, through different working groups.

3.2.1.1 netlmm

The Network-Based Localized Mobility Management (netlmm) working group is developing a two-stage mobility architecture similar from what we know from mobile data networks (e.g., GPRS, CDMA) and from 3GPP's IP Multimedia Subsystem (described in Section 3.3).

Local mobility involves movements across some administratively and geographically contiguous set of subnets, while global mobility involves movements across broader administrative, geographical and topological domains.

During such movements, at least within a geographic area such as a city or campus, the objective is to preserve the higher-layer communication (e.g., a TCP session) by very fast handover and limited packet loss. Other objectives are low signaling overhead.

This work is relevant in a NGN context, as very promising unlicensed access standards (WiMAX) are about to be launched commercially.

3.2.1.2 mipshop, mip4, mip6

The MIP Performance, Signaling, and Handoff Optimization (mipshop) working group has proposed a suitable mobility architecture for low overhead as the Hierarchical Mobile IPv6 (HMIPv6). The standard for fast handover was named FMIPv6.

Previous work in the IETF mip4 and mip6 working groups had focused on the mobility management protocols themselves (MIPv4 and MIPv6), which would allow the devices to announce their current position (real IPv4 or IPv6 address) to a mobility manager in the network. However, these protocols have not known enough commercial success on the IP devices (wireless or wired. cellular or not). In the cellular space for example, MIPv4 was adopted by 3GPP2 CDMA2000 but not by 3GPP GPRS.

On the contrary, commercial WiFi switches appeared to be capable to track a device's mobility across a set of hot spots, *without* specific mobility protocols on the device. The device keeps its allocated IP address, and the handover is performed fast enough for advanced communication applications such as voice not to be affected.

So the IETF netlmm working group is now chartered to standardize an approach that has become a *de facto* market trend.

3.2.1.3 Other Working Groups

Other working groups have addressed the mobility theme. For example, the Seamoby group proposed Candidate Access Router Discovery (CARD, experimental RFC 4066) and Context Transfer Protocol (CXTP, experimental RFC 4067) to speed up the handover of authentication, authorization, accounting, PPP and header compression contexts when the Mobile Node is moving or may be moving from one Access Router (AR) to another.

Redirecting the incoming IP packets as fast as possible to the new AR may not be the sole goal of a mobility management solution.

Of course, the Internet Area includes multiple other working groups aiming for the support of IP (v6 and/or v4) over new physical layers such as 802.16 WiMAX, low power Wireless Personal Access Networks (WPAN) or Digital Video Broadcast (DVB).

Tunneling, for example for remote access to enterprise intranets, also remains a continuous point of attention.

3.2.2 Real-Time Applications and Infrastructure Area — Media Layer

Within this area, the Audio/Video Transport (AVT) working group has delivered the cornerstone for media transport in the NGN: the Real Time Protocol (RTP, IETF RFC 3550, which superseded RFC 1889).

RTP implements application-level framing and integrated layer processing, as defined in David D. Clark and David L. Tennenhouse's paper, "Architectural Considerations for a New Generation of Protocols," ACM, 1990. The sequence numbers in RTP allow the receiver to reconstruct the sender's packet sequence, or the location of a packet in a video stream. RTP supports underlying transport layers with possible packet loss and out-of-sequence arrival (UDP).

Embedded in RFC 3550 is the Real Time Control Protocol (RTCP), of which the aims are to monitor the quality of service and to convey information about the participants in an on-going session. Indeed, RTP and RTCP were designed to make use of IP Multicast at the lower layer. Hence for example, each participant to an audio conference will periodically multicast a reception report plus the name of its user on the RTCP control port.

RFC 3550 also introduces two important RTP relay nodes: mixers and translators. Mixers are able to convert media (RTP packets) from multiple sources into a new outgoing RTP stream, with its own timing. Translators are able to receive multicast RTP packets and send out unicast RTP, for example, to go through a company firewall.

RFC 3550 is not a final protocol specification and makes no assumption about the media stream being conveyed; for each new media type, RFC 3550 needs to be completed with two companion documents:

■ "RTP Profile": the RFC 3551 for audio and video.
■ "RTP Payload Format": how, for example, H.261-encoded video is carried in an RTP packet.

RTCP was extended in 2003 with the eXtended Report (XR) message (RFC 3611), allowing intermediates and end users to discover new objective metrics for media quality: packet loss and discard, burst, delay, Mean Opinion Scores, and so on.

3.2.3 Real-Time Applications and Infrastructure Area — Session Control Layer

3.2.3.1 mmusic

Here we enter the work of the Multiparty Multimedia Session Control (mmusic) working group. I would consider them to be the inventors of IP television.

The Real Time Streaming Protocol (RTSP, IETF RFC 2326, 1998) is sometimes described as "the VCR remote control of multimedia networks." Very similar to HTTP, this protocol allows participants to setup, play, pause and teardown media streams within a session. It is used by the major video players today (RealPlayer, Windows Media Player and Quicktime). The group drafted RTSP 2.0, known as RFC 2326bis, and which can be regarded as a correction of RTSP 1.0.

Immediately next, mmusic catered for a way to describe the media present in a streaming session — the Session Description Protocol (SDP, RFC 2327) and the way these media can be negotiated during session establishment, using an offer/answer model (RFC 3264). In 2006, RFC 2327 was superseded by RFC 4566.

Next, mmusic is preparing a way for clients to announce not only their media capabilities, but also their capabilities to negotiate the media which will be used in a session.

Mmusic is also chartered to deliver an Internet Media Guide (IMG): a collection of multimedia session descriptions, comparable to a television program guide.

3.2.3.2 SIP/SIPPING

This brings us to the work of the Session Initiation Protocol (SIP) working group, in fact a spin-off from mmusic, and the SIP Investigation (SIPPING) working group, which considers new services and feeds requirements into sip.

IETF's cornerstone specification in the session control layer is the Session Initiation Protocol (SIP, RFC 3261), complemented with a procedural RFC on how to extend it (RFC 3427).

SIP is built on a comparable syntax to HTTP, SMTP, RTSP and other IETF signaling protocols.

Whereas RTSP's primary focus was to set up RTP media streams to IP multicast addresses and make it more suitable for broadcasting, SIP creates an infrastructure of clients and servers, comparable to a telephony network, even though IP multicast sessions should also be set up using SIP.

With HTTP, the client (browser) always issues GET or POST requests. With SMTP, the client uses SEND to submit an outgoing e-mail to a SMTP server. With RTSP, the client is the only one issuing the PLAY request.

However, SIP user agents (UA) are able to register to the network previously (using the SIP REGISTER method). Actually the UA consists of a User Agent Client (UAC) side and a User Agent Server (UAS) side, and it's the UAC performing that registration. Following the registration, the network (SIP Proxy server, UAC) can direct incoming session requests (SIP INVITE), messages (SIP MESSAGE) or capability discovery requests (SIP OPTIONS) from the entire world to the destination UA (more specifically its UAS side).

As RFC 3261 says, SIP supports personal mobility: users can maintain a single externally visible identifier (e.g., sip:32496127638@ngvas.com) regardless of their network location. The managers of this mobility are the SIP servers of the domain of that identifier (e.g., sip1.ngvas.com and sip2.ngvas.com). External sources will find these SIP servers by querying the public Domain Name Server (DNS) infrastructure, as explained in RFC 3263 (Locating SIP servers).

The IETF's enum group went as far as submitting a DNS directory structure to IANA for the e164.arpa domain (e.g., 8.3.6.7.2.1.6.9.4.2.3.e164.arpa), in an attempt to let SIP servers discover the next hop to reach any telephone number on the planet. Some countries (e.g., Austria) tried to launch a public ENUM infrastructure. But this attempt to attract the E.164 numbering plan into the Internet has not been very successful, for security, performance, and organizational reasons — who would have had the authority to update this DNS information? To allocate number ranges? Worse, to deal with number portability processes?

Of course, the originating UA (UAC) can also submit outbound requests to a SIP server (UAS; e.g., sip.ngvas.com), in order to reach the rest of the world. That UAS could then react as one of the following:

■ SIP Redirect server: uses 301/203 responses to indicate a new URI (Universal Resource Identifier; e.g., sip:32496127638@ngn.bics.com) to the UAC, asking the UAC send the request to the SIP servers of that new domain.

- Stateless Proxy: will add itself as a Via: line to the header, will proxy the SIP method onwards, receive all nonfinal and final responses (e.g., 200 OK), but not be in the signaling path of subsequent signaling messages of the session (i.e., the forward SIP ACK)
- Stateful Proxy: will adopt the stateless proxy behavior, but also add itself as a Record-route: line in the header, in order to be in the signaling path of subsequent signaling messages pertaining to the same session (ACK, BYE, CANCEL requests and associated responses)
- Back-to-Back User Agent (B2BUA): will act as a User Agent (i.e., negotiate media streams) on either side, hence is capable of complex leg manipulation operations
- Session Border Controller: will also negotiate the media streams using SDP, and force any media (RTP) to be routed through itself, for the following purposes: security, service reach maximization (Network Address Translation traversal), SLA assurance, revenue and profit protection, and regulatory and law enforcement
- SIP Forking server: will proxy the outbound request onwards to multiple destinations — which includes a known security risk for loops (amplification vulnerability, IETF draft-ietf-sip-fork-loop-fix-01)

The IETF sip working group's charter indicates the following next objectives:

- Mechanisms for secure expression of identity in requests and responses
- Mechanism to securely request services delivery by nonterminal elements ("end-to-middle")
- Guidelines for use of existing security mechanisms such as TLS, IPsec, and certificates
- Guidelines for the use of descriptive techniques such as SAML (Security Assertion Markup Language) with SIP
- Draft standard versions of SIP and critical supporting specifications

3.2.3.3 SIMPLE

SIP for Instant Messaging and Presence Leveraging Extensions, that's the description IETF came up with to match the acronym SIMPLE.

This working group is extending SIP to deal with a full suite of services in the domain of instant messaging and presence, in accordance with IETF RFC 2779, the Common Profile for Presence (CPP, RFC 3859) and for Instant Messaging (CPIM, RFC 3860).

The latter used to be managed by the concluded IETF impp working group, dealing with instant messaging and presence protocols (XMPP protocol, RFCs 3920 to 3923). XMPP is today a fairly mature standard, correctly dealing with handshaking, privacy and security, and adopted by the open source community (Jabber) and several IM clients (Google Talk).

The IETF simple group also came up with the XML Configuration Access Protocol (XCAP) used to add/query/remove entries in a buddy list.

The buddy list itself had been created by IETF impp as an XML document named the Presence Information Data Format (PIDF, RFC 3863). SIMPLE inherited the PIDF from the impp, but extended it to cater for watcher information, contact information, rich presence extensions and timed status information.

3.2.4　Security Area

Next to mobility aspects, security is a rising general concern for new communication protocols and architectures. The success of a technology like GSM is partly due to the facts that nobody ever succeeded to break GSM authentication (i.e., to impersonate someone else's Subscriber Identity Module) nor to pull down a GSM network with a Denial-of-Service attack.

Current Internet Protocol security protocol (IPsec), Transport Layer Security (TLS) and Internet Key Exchange protocol (IKE) present somewhat of an all-or-nothing alternative; these protocols provide protection from a wide array of possible threats, but, as the Better-Than-Nothing Security (BTNS) working group recognizes, are sometimes not deployed because of the need for preexisting credentials, such as the ones inside a GSM SIM.

3.2.5　Conclusion for the IETF NGN

The IETF continues to deliver, by far, the most solid standardization work for the NGN. It could benefit from a more coherent approach toward mobility management, though.

At Network level, the IETF promoted device-impacting technologies such as IPv6, MIPv4/v6, HMIP/FMIP, and, in general, IP over new bearers. It now realizes that it could have decoupled the non-device-impacting mobility inside a local area, from whatever mobility protocol in the core of the Internet (e.g., MIPv4, MIPv6, 3GPP GPRS Tunneling Protocol, SIP REGISTER).

In the Media and Border Gateways layer, IP multicast and RTP are brilliant technologies but the generally perceived VoIP quality is low on the public Internet, except with QoS enforcement or peer-to-peer protocols.

In the Session Control layer, the IETF also came to the conclusion that mobility needs to be hierarchical, namely as hierarchical as there will be SIP proxies in the path of an incoming SIP INVITE. In theory, IETF's DNS queries are going to be slower than 3GPP's DIAMETER queries to a Home Subscriber Server (HSS).

But the degree of mobility communication consumers are expecting is rapidly increasing anyway: cross-network, cross-technology and with seamless in-session handover.

This brings us to the work of the 3GPP.

3.3 3GPP

The 3rd Generation Partnership Project (3GPP) is the current owner, from network standards perspective, of GSM (2G*), the packet-based GPRS (2.5G†), and related higher-speed cellular mobile wireless technology (2.75G‡–3G§–3.5G¶). 3GPP inherited the responsibility for 2G technology from ETSI SMG (Special Mobile Group).

Some of the 3GPP technologies are pure radio network technologies (EDGE, UMTS, HSPA), designed to enhance access speeds (for packetized data) in the radio access network, with as little impact as possible on the core network. In fact there's only one technology for the packet-based core network: GPRS.

It is on top of this packet-switched domain that 3GPP defined its "NGN plus" as the IP Multimedia Subsystem (IMS).

3GPP's strategy is one of Fixed-Mobile Convergence (FMC), with as midterm target that the IMS would become the umbrella core network not only above the mobile packet-switched radio access network but also above the fixed broadband access networks (e.g., DSL, cable modems, DVB). The IMS would offer solutions for universal problems such as strong authentication, mobility management across access networks, and quality-of-service over packet networks.

The assumption is that the mobile and fixed networks in a same country will be run by a same administrative entity, offering multiplay services.

The services layer, the supervision, provisioning and charging/billing infrastructures could then be shared among fixed broadband and mobile access divisions, producing cost savings in these FMC-ready organizations.

3.3.1 UMA

Step zero of FMC has been the adoption by 3GPP of Unlicensed Mobile Access (UMA) technology, where the WiFi hot spot federation needs to be directly connected to the GSM Network SubSystem (via the so-called A-interface) for voice, and to the Packet-Switched Core (via Gb/Iu interface) for data. Thus, UMA access could not be offered from say a hotel's WiFi hot spot in a foreign visited country, as the hotel's ISP is not connected to a GSM/GPRS core network.

UMA is an extension of the 2G-2.5G authentication and *local* (access network) mobility schemes to cover the WiFi domain.

Due to the immaturity of IETF's SIP technology for authentication and mobility management issues, 3GPP was forced to adopt the GSM A-interface protocols (BSSMAP/DTAP, 3GPP TS 04.08) and to rely on the GSM SIM.

---

* Global System for Mobile communications, the second generation cellular network.
† General Packet Radio Service.
‡ EDGE: Enhanced Data rates for GSM Evolution.
§ UMTS: Universal Mobile Telephony System, also known as Wideband CDMA (W-CDMA).
¶ HSDPA: High Speed Downlink Packet Access.

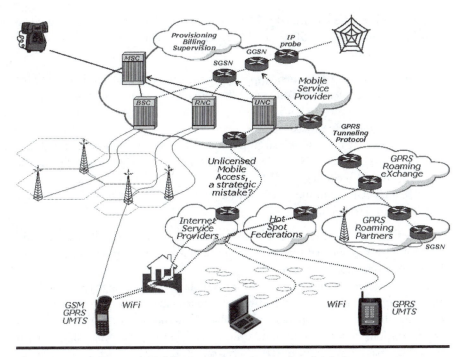

Figure 3.2   Unlicensed Mobile Access — a strategic mistake?

UMA is probably a historic missed opportunity to extend the 2.5G/3G global mobility management (GPRS/UMTS core network) toward independent, visited federations of WiFi hot spots, connected to the GPRS Roaming eXchange (GRX) backbone via the GPRS Tunneling Protocol (GTP, 3GPP TS 09.60 / 29.060). Some independent companies* had built these bridges between hot spot federations and the GRX providers, but mobile service providers, having ambitions to roll out their own hot spots in the homes, never adopted this principle.

The local mobility protocol (across the hot spots of a single federation) would then have been out of scope of 3GPP. It could have been IETF MIPv4, MIPv6, a proprietary non-device-impacting solution, or better, the upcoming non-device-impacting work of IETF netlmm.

Authentication could have used Extensible Authentication Protocol in the access network (EAP, IETF RFC 3748) and Mobile Application Part in the core (MAP, 3GPP TS 09.02 / 29.002). Or SMS authentication, a very strong scheme in which the mobile phone owner requests a password to be sent by SMS to the mobile handset, and enters his mobile phone number plus that password to access the hot spot (or service in general). Or any other mechanism to verify an existing active charging/billing relationship with the home mobile operator.

---

* That is, the WeRoam.com offer by Comfone of Switzerland, part of the Togewa holding.

Some will see the hand of the GSM Association, who would have refused to grant full membership to ISPs and WiFi federations, therefore would have denied them the rights to

> Close roaming agreements, specifying their inter-operator tariffs (IOT) for Internet access via WiFi and other broadband access technologies
>
> Connect to the international SS7 networks (using MAP for authentication and mobility procedures)
>
> Use CAMEL (Customized Application of Mobile Enhanced Logic, 3GPP TS 03.78 / 23.078 / 09.78 / 29.078) for real-time charging to the mobile prepaid account
>
> Connect to roaming clearing houses, for inter-operator CDR exchange and settlement

The aforementioned scheme (let's call it UMA 2.0) was actually realized and is currently in commercial operation by a few hot spot federations owned by, or closely linked to, mobile operators.

From a macroeconomic perspective, it would have made sense to reuse the existing charging, top-up, billing and customer care resources (infrastructure, employees) of the (home) mobile operators. Broadband Internet service providers and hot spot federations could have become diligent and loyal resellers of what remains essentially a basket of mobile operator services.

It is also a safe bet that the addition of unlicensed Internet access would have increased the general adoption and turnover of mobile data.

But UMA 2.0, by supporting VoIP and IM, would have put circuit-switched voice and sms revenue and price under strong pressure. It's a battle the GSM operators aren't prepared to fight. Today's regulatory threats to roaming revenues are a little teaser compared to what could happen in case of widespread UMA 2.0.

UMA (1.0) market trials have indeed shown that in practice, 70% of communications were made through the residential hot spot, and 30% via GSM.

3.3.2 IMS

Fundamentally, the 3GPP IMS's goal is to be an IETF NGN "plus."

But, as we will see in the upcoming Section 3.5, it soon became a TISPAN NGN "minus."

3GPP has used its recognized experience in the domains of authentication and mobility to address the main "holes" in the IETF NGN: security, mobility, service hooks and QoS enforcement. These are actually area's which are still open, constantly evolving and where it's not sure whether a dominant solution will emerge. Let's therefore consider them more closely in the upcoming sections.

Besides these topics, 3GPP IMS also

> Defined private and public user identities

Extended SIP (in IMS Stage 3, TS 24.229 Release 7, Section 7), and adopted
   SIP signaling (header) compression (in Section 8)
Specified the interworking of the IM core network with the circuit-switched
   domains (GSM/UMTS/PSTN/ISDN) using ISUP and BICC (in TS
   29.163 Release 7)
Embraced some services such as presence, instant messaging, conferencing,
   group management

It should be noticed that 3GPP uses slightly different terminology for the functional planes:

"Control Plane" for session control
"User Plane" for the border and media gateways

3.3.2.1 Security

Whereas the distribution of authentication credentials is not addressed in an IETF
NGN, 3GPP specified an own access security mechanism (AKA: Authentication
and Key Agreement, TS 33.203) based on SIP registration and on a secret number
in the ISIM, rather than on IETF's lower layer protocols (IPsec/TLS/IKE). The
IMS User Entity (UE) thus consists of a SIP User Agent (UA) and the IP Multimedia Subscriber Identity Module (ISIM).

Maybe 3GPP should have been more forward-looking and adopted the *Extensible Authentication Protocol* (EAP, RFC 3748) instead of SIP REGISTER.

The security algorithm itself, which the ISIM uses to calculate a response
(XRES) for a random challenge, with the ISIM's secret number as the other input,
is not part of standardization.

Good access security also means that the UE should also be able to authenticate
the IMS. Therefore, the UE receives a message authentication code (MAC) from
the network, and calculates an expected message authentication code (XMAC) to
see if these match. The algorithm is also out of scope of standardization.

Once the ISIM has been authenticated by the IMS, and vice versa, SIP signaling
between the User Agent and the network is encrypted using an agreed session key.

In an IMS, the IETF's lower layer security protocols are only used for network
domain security (between the SIP servers and HSS, see 3GPP TS 33.210).

3.3.2.2 Mobility

Contrary to the IETF which is still struggling, ETSI and 3GPP have always realized the benefits of hierarchical mobility architectures. The structure of the GSM
radio and core networks reflects this:

The Base Transceiver Station (BTS) preserves radio contact while the
   Mobile Station (MS) moves within one or a few cells.
The Base Station Controller (BSC) manages the mobility of the MS across
   a set of BTS.

The Visited Mobile Switching Center (VMSC) manages the mobility of the MS across multiple BSCs. When the MS moves from VMSC to another, the Home Location Register (HLR) keeps track of where to find the MS (i.e., at which VMSC), should there be incoming calls or messages. If the new VMSC is within the same network (i.e., the MS is not roaming out to a foreign VMSC), ongoing calls are seamlessly handed over from the old to the new VMSC (but the old VMSC remains involved in the call).

The Gateway Mobile Switching Center (GMSC) is able to direct inbound calls to the correct VMSC, by querying the HLR (the HLR then obtains a temporary number from the VMSC). For incoming calls, the GMSC and HLR thus manage the top level mobility in a 2G network. The same design philosophy is present in the Packet-Switched domain (2.5G–2.75G–3G–3.5G PS) and in the IMS (see 3GPP 23.228, IMS Stage 2).

The node-B preserves radio contact while the User Entity (UE) moves within one or a few cells.

The Radio Network Controller (RNC) and Packet Coding Unit (PCU) manage the mobility of the UE within a Routing Area (i.e., a set of cells).

The Serving GPRS Support Node (SGSN) manages the mobility of the UE across multiple Routing Area's. When the MS moves from SGSN to another, the Home Subscriber Server (HSS) keeps track of where to find the MS, should there be network-requested Packet Data Protocol (PDP) contexts. If the new SGSN is within the same network (i.e., the MS is not roaming out to a foreign SGSN), active PDP contexts are seamlessly handed over from the old to the new SGSN.

The Gateway GPRS Support Node (GGSN) forms the permanent point of interconnection with Public Data Networks (PDN) such as the Internet. PDP contexts are created between the SGSN and GGSN using a protocol called GTP (GPRS Tunneling Protocol). The GGSN is informed by GTP if the UE moves to a new SGSN while a PDP context is active. The GGSN is also capable to query the HSS to find the UE's current SGSN, for network-requested PDP context activation toward that UE.

A Proxy Call Session Control Function (P-CSCF) is the first point of contact for all SIP messages between the UE and the IMS; it receives the SIP REGISTER request from the UE, and proxies it to an Interrogating CSCF (I-CSCF), which queries the HSS. The HSS allocates a suitable Serving CSCF (S-CSCF) and the I-CSCF proxies SIP REGISTER to that S-CSCF. The P-CSCF proxies outbound (i.e., UE-originated) SIP INVITE directly to a S-CSCF in the Home Network. The P-CSCF is also the first point of contact for inbound sessions to the Visited Network.

The I-CSCF in the Home Network is capable of querying the HSS in order to find the UE's current S-CSCF, for inbound sessions toward that UE.

The I-CSCF and HSS thus manage the top-level mobility in an IMS, for inbound sessions.

However, the need to break the CSCF into three entities, or even the need for an open interface between CSCF and HSS has gone away, since 3GPP recognized that the IMS would be accessed solely through GGSNs of the Home PLMN.

Indeed, in the early days of IMS (2000-2002), it was believed that a P-CSCF in Visited Network A (just next to Visited GGSN A) would have to communicate with an I-CSCF in Home Network B, via SIP and over an international network called IPX (IP eXchange). Also, a S-CSCF may have been assigned in the Visited Network A (just next to Visited GGSN A), to communicate with the HSS in Home Network B.

Today, we know that in practice, the SGSN A will only communicate to a GGSN B in the Home Network, over the GRX (GPRS Roaming eXchange). Indeed, as many services (e.g., real-time charging, IP content filtering) are Home GGSN-based, using Visited GGSNs is not an option anymore. Moreover, there had been the unsolved questions about how the inbound roamer B would find the Access Point Name to GGSN A in a visited network, why he/she would even try to, and whether that APN would be have been provisioned in his/her HLR B (or just a default setting then allowing any APN).

Hence as all elements (P-CSCF, I-CSCF, S-CSCF, HSS but also SBC and MGCF) are all in the subscriber's home network anyway, the functional separation into different logical entities is not needed anymore, and just makes *inefficient use of resources*, compared to a traditional architecture of Class 5 and Class 4 NGN softswitches. The elements waste a lot of energy (CPU power) to encode/decode messages between each other, rather than spending these cycles on the interfaces to the external world (user entities and other networks).

3.3.2.3 Service Hooks

Besides session-level mobility, 3GPP provides the hooks to let the session plane trigger/invite the service plane, but only at call/session setup stage.

The 2G-2.5G-3G mechanism is CAMEL.* The HLR holds the necessary CAMEL subscription information (CSI, including the address of the Service Control Function). For originating calls, the HLR inserts O-CSI into the Visited Network (Visited MSC/VLR) when the mobile phone is initially powered on (more specifically during the Location Update/Insert Subscriber Data procedure). For terminating calls, the T-CSI is interpreted by the Home Network (Gateway MSC) when received from the HLR (in response to a Send Routing Info request).

The IMS mechanism is the "evaluation of initial filter criteria" and is performed by the S-CSCF in the Originating or in the Terminating Home Network, which can invite SIP Application Servers (AS). Again the HSS holds the necessary criteria.

The session plane needs to trigger/invite important "originating" services such as prepaid/real-time charging, lawful interception, and centrex. But increasingly,

---

* Customised Application of Mobile Enhance Logic.

the session plane will need to trigger and invite services regarding terminating calls and sessions.

A destination user may indeed have chosen to receive the inbound call/session in a non-GSM or non-IMS environment, meaning without having informed the HLR/HSS about its location in that new environment. Maybe the user has devices located simultaneously in both 3GPP and non-3GPP networks.

Examples include the following:

> An instant messaging, gaming user, or IPTV watcher wishing to receive inbound calls, messages and sessions inside an IM program, online game or MPEG2 stream, when logged on there (rather than to the cell phone)
>
> A business user away from his/her desktop PC, and expecting the incoming calls to be handled by the company's receptionist
>
> The mobile user of a standard dual-mode GSM/WiFi phone, supporting VoIP and maybe SIP, but not necessarily IMS, and expecting seamless WiFi-to-GSM handover
>
> An outbound roamer, for whom it is cheaper to receive incoming VoIP calls on a plain vanilla VoIP/WiFi handset than on the roaming GSM phone
>
> An American mobile user, who pays for incoming calls to the mobile side but not for incoming VoIP calls through a WiFi hotspot

With revenues from subscriptions, outbound, and international roaming traffic decreasing, it becomes very important for 3GPP network operators to attract more inbound traffic through its GMSCs and I-CSCFs as the anchor points.

Though 3GPP networks would support these terminating services (as well as of course the originating) in a basic form, the hooks in the session plane are insufficient to compete with nonstandard implementations, hence to keep attracting that inbound traffic.

When interrogated by the I-CSCF in the Terminating Home Network, a first hurdle is that the HSS should accept to reply with a default S-CSCF, if the destination subscriber is not registered in the HSS (or even unknown). It is a design flaw that the I-CSCF cannot query the SIP AS *directly* (as the GMSC can query the CSE*).

The service plane can then instruct the session plane to proxy the call or session to the non-3GPP environment, via a conversion gateway.

Second, 3GPP GSM and IMS offer no hooks in the call/session plane to provide seamless, *in-call handover* from one environment (e.g., inbound VoIP over WiFi) to the other (e.g., GSM). This is true for both terminating calls/sessions to the E.164 telephone number, as for calls/sessions originating in a 3GPP network which need to be handed over to a non-3GPP environment.

---

* CAMEL Service Environment.

The key decision when to perform that handover can be made only by the service plane, based on presence, availability or VoIP quality measurements. But the service plane would then need to instruct the session plane "mid-call"/"mid-session."

Third, can the 3GPP IMS session plane invite the service *selectively*, only for certain media types, origins or destinations (e.g., not for emergency calls)?

Next, is the 3GPP IMS session plane going to be capable of inviting *multiple services* (for example, prepaid and lawful interception) for a given session, in parallel or sequentially? Or will the SIP AS have to proxy one to another? It's too easy for 3GPP to push out the service orchestration issues to the service plane.

In summary, 3GPP networks do support some hooks to the service plane, but will need to progress in this area.

For inbound traffic, communication service subscribers today demand added value: the convenience to be reachable behind an expensive E.164 mobile phone number is not enough anymore.

If 3GPP fails to enable the new FMC and multiplay services, in the near future we might well set up our calls and sessions using SIP to e-mail addresses instead of to E.164 telephone numbers.

3.3.2.4 QoS Enforcement

In the 3GPP IMS, the P-CSCF contains a Policy Decision Function (PDF), which can instruct the access network node (Gateway GPRS Support Node, Mobile IP Home Agent or Broadband Remote Access Server) to apply a given IP Quality of Service (IP QoS) via Common Open Policy Service (COPS) protocol on the Go interface. This is the COPS response to an initial query by the GGSN.*

In order to understand the IMS session binding and QoS enforcement, let's consider the case of an IMS session being set up between two User Entities (UE 1 and UE 2). UE 1 is registered to the Originating IMS 1, and UE 2 to the Terminating IMS 2. Therefore, a primary access channel (e.g., PDP context, PPPoE session) must have been created to convey the SIP signaling in IMS 1.

Upon SIP INVITE, the Originating S-CSCF as well as the Terminating S-CSCF will verify that UE 1 and UE 2 are entitled to set up the media streams proposed by UE 1 (described in the SDP, RFC 2327). Indeed, the HSS contains the allowed media.

SIP INVITE is acknowledged hop-by-hop with a "100 Trying" provisional response, in order to avoid time-outs.

The PDF at the Terminating P-CSCF produces an IMS Authorization Token and adds it in the SIP INVITE toward UE 2.

---

* And, in the future, by a Home Agent (HA) for Mobile IP networks, or a Broadband Remote Access Server (BRAS) for DSL networks.

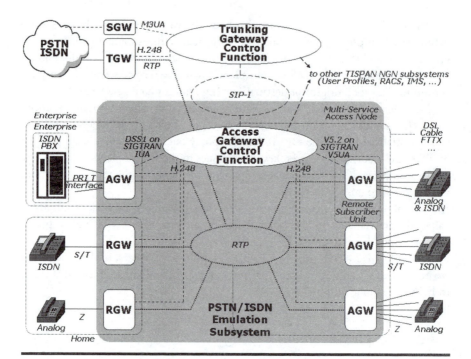

Figure 3.3   Session setup and quality-of-service enforcement in IMS.

UE 2 communicates its SDP information (according to RFC 3264) in a backward SIP 183 Session Progress message, which will be authorized by both S-CSCF.

Also at this point, UE 2 already generates Flow-IDs, for each secondary access channel (PDP context) that will be set-up for a media type listed in the SDP. The PDF at the Terminating P-CSCF in IMS 2 authorizes the use of these resources: "is there enough overall bandwidth left to set up all these secondary access channels?"

So does the PDF at the Originating P-CSCF in IMS 1, upon receipt of SIP 183 Session Progress. It also produces an Authorization Code.

Upon receipt of the SIP 183 message, UE 1 produces Flow-IDs for the Authorization Token, and informs UE 2 (SIP PRACK).

Now UE 2 activates secondary access channels for each of the media types involved, passing the combination of Authorization Token and Flow-ID. While doing so, UE 2 requests an appropriate access QoS for each media type listed in the SDP. Strangely, in IMS, *it is up to the UE to select the access QoS*, so it could request a high UMTS QoS for media not requiring it!

The local GGSN verifies the bindings between Authorization Token and Flow-IDs, via COPS protocol on the Go interface. Go is set up between a Policy Control

Function (PCF, terminating GGSN in this case) and the PDF (at the Terminating P-CSCF). "Yes, these resources have been previously authorized ..."

During this process, the PDF informs the PCF about the maximum IP QoS to be applied for each access channel. The PCF will also inform the PDF/P-CSCF about the Charging Correlation Identifiers (IMS Charging ID, GPRS Charging ID) which will allow correlating duration-based IMS charging with volume-based access charging. When a time-based IMS session took place, the corresponding access volume charges will probably be waived.

When all secondary access channels are successfully set up, and the bindings verified, UE 2 informs UE 1 by acknowledging the PRACK message (with SIP 200 OK).

UE 1, the Originating GGSN and PDF/P-CSCF in IMS 1 now also set up the secondary access channels and verify the bindings.

At the end, SIP UPDATE informs IMS 2 and UE 2, and when UE 2 acknowledges (SIP 200 OK), the QoS is finally enforced.

Next (not shown anymore on the diagram), IMS 2 will send a backwards SIP 180 Ringing signal, which will be acknowledged by IMS 1 with a forward PRACK (and IMS 2 acknowledges that PRACK with backwards 200 OK).

When UE 2 finally accepts the session (takes the call), a backwards 200 OK message flows to UE 1, which acknowledges that to UE 2 (forward SIP ACK). Media can then finally flow between UE 1 and UE 2.

The IP QoS can be of the DiffServ type, where access network router sets the IP Type of Service (ToS) bits, also called DiffServ Code Point (DSCP, IETF RFC 2427). DiffServ is a coarse QoS mechanism, as the same QoS is then applied for all packets in that access channel (e.g., PDP context)

Optionally in some PCF, the IP QoS can be of the IntServ type, a fine-grained QoS mechanism by which the access node (e.g., GGSN) will use Reservation Protocol (RSVP) to request a QoS better than best effort, for each uplink and downlink media stream within a single access channel (e.g., PDP context).

IMS QoS enforcement not only impacts the access network nodes (e.g., GGSN) but *also the User Entities* of each present and future access technology. That is, it requires new mobile phones, which link the SIP layer to the transport technology (e.g., GPRS, UMTS).

As such, IMS QoS enforcement may actually never happen.

Seen in the broad picture of a User Plane, a Session Plane and an Service Plane, again is difficult to see how the Service Plane will:

> Be informed about the proposed and the ultimately negotiated media streams (e.g., for prepaid tariff determination)
>
> Influence the QoS being applied (e.g., increase the QoS for calls to/from the operator's own IP access networks)

So, last but not least, QoS enforcement in IMS is *service-agnostic*.

3.3.2.5 Conclusion for the IMS

Regarding the IMS, there are some concerns that:

> The degree of security, mobility and service triggering is not sufficient to justify the investment in a much more complex system than the IETF NGN.
>
> The decomposed architecture itself of the IMS (breakdown of a softswitch to P-CSCF, S-CSCF, I-CSCF and even HSS) cannot be justified anymore, and wastes computing resources on internal interfaces.

QoS enforcement is complex in IMS, increases the delay to set up a basic call, and requires close interaction between the SIP signaling and underlying mobile data transport layer in the terminal; it is thus not possible to just download an IMS User Entity on any IP device.

> The features of PSTN/ISDN networks (supplementary services, and regulatory obligations such as lawful interception and number portability) will not be supported by the IMS itself anyway.
>
> Messaging (SMS, MMS) appears to have been totally forgotten in the IMS.
>
> Architectural separation of a session plane and a user plane is possible, even in the circuit-switched (CS) domain of 2G mobile networks, using MSC servers controlling media gateways, without deploying an IMS on the packet-switched (PS) side.

3GPP stretched itself from the mobile/wireless space into fixed broadband, an area with very different requirements.

An IETF NGN would outperform a 3GPP IMS if that IETF NGN would be augmented with:

> User authentication via EAP or SMS (instead of ISIM/USIM).
>
> TLS/PKI security between network nodes (same as in IMS).
>
> Non-device impacting local mobility (instead of 2.5G-3G mobility management of the GERAN/UTRAN).
>
> Either 3GPP GTP, IETF SIP or IETF MIP as global mobility mechanism (instead of 3GPP GTP + IMS SIP), with header compression (as in IMS).
>
> No QoS enforcement (best effort service), but downgraded DiffServ QoS for peer-to-peer services.
>
> Dedicated SIP servers, proxying to each other.
>
> Session Border Controllers on the access and interconnect sides (same as in IMS).
>
> Class 4 Media Gateway Control Function (supporting SS7 ISUP, SIP-T and H.248) and Media Gateways (same as in IMS).
>
> Class 5 Softswitch for PSTN/ISDN emulation.
>
> IP Centrex Server for enterprise PBX services (perhaps asterisk, perhaps still with H.323 and ISDN support).

SIP AS for real-time charging (in Stateful Proxy mode).
FMC server for voice call continuity (in B2BUA mode).
Media Server, controllable by VoiceXML and SIP.
An XMPP or SIP/SIMPLE IM and presence server.
An SMS/MMS gateway.

Some of these items are today addressed by TISPAN, of which we'll review the work in Section 3.5.

3.4 3GPP2

The 3rd Generation Partnership Project 2 was established in 1998, as a cooperation between standards institutes from Japan, China, North America, and South Korea.

3GPP2 is today driving the CDMA2000 family of packet-based technologies, as the successor to 2G CDMA (cdmaOne, TIA IS-95), and in analogy with 3GPP driving the evolution of GSM.

Within the CDMA2000 family, multiple generations of mobile wireless cellular technologies were born: 2.5G-2.75G (CDMA2000 1x) and 3G (CDMA2000 1x EV-DO*).

On top of these radio access technologies, and besides the circuit-switched core network, 3GPP2 developed a packet-based core network similar to 3GPP GPRS. It consists mainly of Packet Data Serving Nodes (PDSN, the equivalent of 3GPP's SGSN) and the Home Agent (HA, the counterpart of 3GPP's GGSN).

Contrary to 3GPP, 3GPP2 adopted IETF's Mobile IP protocol, allowing the HA to track a subscriber's mobility across a set of PDSN.

But on top of this core network, 3GPP probably made a strategic mistake not to adopt the IMS, and ended up specifying their own MultiMedia Domain (MMD).

Also, 3GPP2 does not have the market power to develop an FMC strategy, to let this MMD span fixed access networks too.

It is not the first time that 3GPP2 would define a technology that would stall due to lack of implementation and adoption (e.g., CDMA2000 1x EV-DV†).

3.5 TISPAN

ETSI has bundled its Next Generation Network efforts in the Telecommunications and Internet Converged Services and Protocols for Advanced Networking (TISPAN) "one-stop-shopping" working group, a merger of the former groups:

Network Aspects (NA)
Signaling Protocols and Switching (SPS)

---

* Evolution for Data Only, or "Data Optimized," TIA IS-856.
† Evolution for Data and Voice, of which Qualcomm halted the development in 2005.

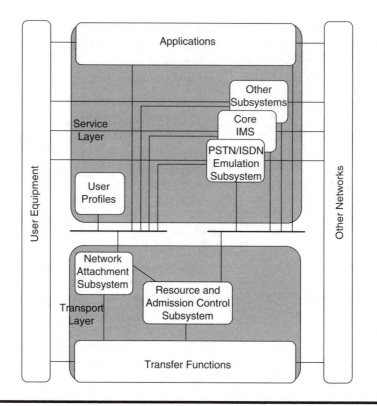

Figure 3.4   TISPAN NGN overall architecture.

Services and Protocols for Advanced Networks (SPAN)

Telecom and Internet Protocol Harmonization Over Networks (TIPHON)

Whereas ETSI's influence was confined to Europe, TISPAN.org is now leading an independent and international life, just like 3GPP.

And we must say, with success.

Officially, TISPAN has incorporated the 3GPP IMS at the heart of its NGN architecture and there will be a single IMS for both mobile and fixed access. The following diagram shows the overall architecture of the TISPAN Next Generation Network, as described in the standard ES 282 001:

In fact, TISPAN has corrected the mobile-centric aspects of 3GPP IMS releases 5 and 6, and many shortcomings in the fixed access part of 3GPP IMS release 7.

TISPAN NGN Release 1 contains over 70 deliverables of various types:

12 Technical Reports (TR)

39 Technical Specifications (TS) issued by TISPAN only

20 ETSI Standards (ES) endorsed by the ETSI members

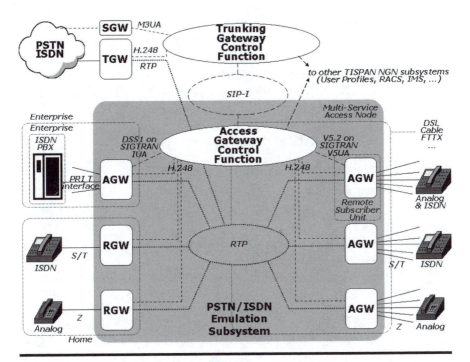

Figure 3.5   TISPAN PSTN Emulation Subsystem (PES).

The specifications cover service requirements, architecture and protocols, plus a few extra aspects such as security, QoS, ENUM and TMN.

TISPAN has endorsed the 3GPP IMS Stage 2 architecture (TS 182 006, 3GPP 23.338), corrected 3GPP Release 7 IMS for fixed broadband access (ES 282 007), and added the

> PSTN/ISDN Emulation Subsystem (PES, TS 182 012 and ES 282 002)
> Resource and Admission Control Subsystem (RACS, ES 282 003)
> Network Attachment SubSystem (NASS, ES 282 004)
> Emergency Call Services (ECS, TS 182 009)

Figure 3.5 represents the architecture of the PES:

This diagram is inspired by ES 282 002, but illustrates the position of the various gateways in the home, the enterprise and the service provider realm.

Within the PES, the Access Gateway Control Function (AGCF) has interfaces

> To the Residential GateWays (RGW in the home), controlled via ITU-T's H.248 protocol, and offering local analog telephony (twisted copper pair at the Z reference point) or ISDN (double copper pair, S/T interface point).

To ISDN Access GateWays (AGW) located within the enterprise, or on service provider premises, offering the ISDN Primary Rate Interface (PRI service at the so-called "T" reference point) and more advanced variants (DPNSS, QSIG). The AGCF receives ("backhauls") the ISDN DSS1 signaling over SIGTRAN ISDN User Adaptation (IUA), and pilots the AGW via H.248 protocol, in order to create RTP media streams to the IP access network.

To V5.2 Access GateWays (AGW), which were referred to as "Remote Subscriber Units" (RSU) in classic telephony, and which are connected to the core network via the V5.2 interface: the V5.2 protocol over SIGTRAN V5 User Adaptation (V5UA). The AGCF also pilots these AGW via H.248.

To analog and ISDN Access GateWays (AGW) also controlled via H.248, within a Multi-Service Access Node (MSAN). A MSAN may include V5/RSU technology, and will also provide the broadband Internet access services (using DSL technologies, cable, fiber to the home, etc.). Again, the AGCF pilots the AGW via H.248.

To the SIP-I core network (SIP for Interconnection, see Section 3.6.5), for example, in order to reach the Trunking Gateway Control Function (TGCF). The TGCF controls Trunking GateWays (TGW) via H.248, and sends SS7 ISUP over SIGTRAN M3UA (or MTP3 over SIGTRAN M2PA) via Signaling GateWays (SGW) to the legacy PSTN and ISDN.

To the other TISPAN NGN subsystems, which includes the RACS. The RACS deals with admission control (possibly with overbooking), resource reservation (for bearer resources such as port numbers), policy control (to enforce IP QoS), remote NAT traversal and local NAT. It is composed of

The Service Policy Decision Function (SPDF): a single point of contact for Access Functions (AF) such as the Proxy-Call Session Control Function (P-CSCF) or Interconnection Border Control Function (I-BCF) over a DIAMETER Gq' interface; the SPDF opens and closes the gates for communication on the Core or Interconnection Border Gateway Functions (C-BGF or I-BGF).

The Access and Resource Allocation Control Function (A-RACF): is invoked by the SPDF in order to take decisions based on current network load, and communicates them to the IP Edge, for example, a Broadband Remote Access Server (BRAS) for xDSL access (Layer 2 Termination Function and Resource Control Enforcement Function), via the Re interface.

The NASS is responsible for IP address allocation and other traditional Authentication, Authorization and Accounting (AAA) functions in the access network. It is invoked by an Access Management Function (AMF) at the IP edge (i.e., the BRAS, in the xDSL access example). It stores the IP address which was allocated to a given user device, as this may be useful information to applications.

The TISPAN architecture is very complete and advanced, however one may wonder whether it doesn't require specialized entities or sub-contractors to operate the RACS and NASS.

The TISPAN NGN is much more versatile than the "naked" 3GPP IMS. It can be used to provide PSTN emulation services, SIP-based IMS services, IP TV, peer-to-peer solutions, or even allows the end user to request more bandwidth on the fly — sometimes referred to as "the Turbo button."

3.6 ITU-T

Throughout the 1980s and the 1990s, the Telecommunications Standardization Sector of the International Telecommunications Union (ITU) produced the most important standards for public and private circuit-switched telecommunications.

3.6.1 ISDN

Digital telephony was introduced at the end of the 1980s under ITU-T's ISDN standards (Integrated Services Digital Network): the Q.9XX series of recommendations.

The ISDN spanned the digital core network, the residential access network (analog and digital), and the enterprise network (digital PBX) — as the NGN today does.

British Telecom completed ISDN with DPNSS (Digital Private Network Signaling System), allowing advanced supplementary services to be supported across multiple company PBX's.

ETSI took a similar initiative (QSIG) which was more widely adopted, but DPNSS remains technically superior.

3.6.2 SS7 Signaling

In order to convey the core network signaling, ITU-T created Signaling System number 7 (SS7) as the Q.7XX series of recommendations, with a first stable version in 1988 (the so-called Blue Book) and quite stable revision in 1993 (the White Book). It has permitted national and international interconnection of digital telephony exchanges. It is universally appreciated for its accuracy, robustness, excellent interworking (e.g., with international ISUP, Q.767). Even though inevitably, regional variants have appeared (ANSI SS7).

SS7 is structured as the Message Transfer Part (MTP) at data link and network layer, and either the ISDN User Part (ISUP, for call set-up), or the Signaling Connection and Control Part (SCCP, for other signaling). Above SCCP, the Transaction Capabilities Application Part (TCAP) allows associate requests and responses in a dialogue. (Wow! This is probably the shortest overview ever on SS7!).

The Global System for Mobile communications (GSM), the largest telecommunication system on earth, relies heavily on SS7, for call set-up as well as for mobility tracking (location update procedures, routing of incoming calls and messages) and security (terminal and SIM authentication). Overall, probably 80% of the signaling exchange for calls, SMS and mobility events uses pure SS7. A small portion in the mobile access network (between MS and MSC, and between BTS and BSC) is based on ISDN access signaling.

An interesting ITU-T development is Bearer Independent Call Control (BICC, Q.1901), a signaling protocol allowing broadband multimedia streams to be set up (Q.1970) besides narrowband circuits to be connected (Q.1950). Also, BICC signaling itself can be conveyed over a broadband network (IP). Besides the two core network protocols, there's also a BICC variant for access signaling (Q.1930).

BICC Capability Set 2 (2001) was adopted by 3GPP for its Release 4 UMTS specifications.

3.6.3 Intelligent Network

Part of the success of GSM is due to the ability for customers to use it without preestablished contract, monthly subscription fee, or billing address. The prepaid mode has dramatically lowered the barrier of entry, and is used today by more than 60% of all mobile phone users worldwide. This fraction is bound to rise further, as almost all new users are connected in developing countries.

With their Q.12XX series laying the basis of the Intelligent Network, ITU-T created the standard to let digital telephony exchanges communicate with external computers (the prepaid IN platforms). Also there, more refined variants have appeared (ETSI Core INAP, GSM CAMEL, CDMA Wireless IN) but the impact of ITU-T has been decisive. Though the IN was foreseen to become a multiapplication VAS platform, the killer app is prepaid.

ITU-T's Intelligent Network Application Part (INAP) is carried over TCAP, over SCCP, over MTP. But it can also be carried over IP networks as INAP over TCAP over SIGTRAN SUA.

3.6.4 Media Encoding

Also in the media and border gateways layer, ITU-T is the main actor, for the ISDN, GSM and NGN. Echo canceling and the compression of audio and voice (multiple codecs) are standardized in ITU-T's G series. The most commonly used codecs for VoIP are

G.711: 64 Kbit/s, also used for circuit-switched digital landline telephony, MOS* = 4.1.

G.729.1: as of 14 Kbit/s, for wideband audio (50–7000 Hz), MOS ≥ 4.

G.729: 8Kbit/s only, but unable to transmit DTMF or fax, MOS = 3.92.

---

* Mean Opinion Score, a subjective average mark given by listeners on a scale between 1 and 5.

G.729a: 8 Kbit/s, less computing-intensive version of G.729, MOS = 3.7
G.723.1: 6.3 or 5.3 Kbit/s, MOS = 3.9 or 3.65

In noisy wireless environments, ETSI and 3GPP have been in the driving seat:

GSM-FR: Full Rate, 13 Kbit/s, invented in the early 1990s but today out-dated, MOS up to 3.2.

GSM-HR: Half Rate, 6.5 Kbit/s, also from the early 1990s, MOS < 3.

GSM-EFR: Enhanced Full Rate, 12.2 Kbit/s, delivers wireline-like MOS up to 4.5 in a quiet street (4.1 in-car with 10^{-3} BER*).

AMR: Adaptive Multi Rate, from 4.75 to 12.2 Kbit/s, adopted by 3GPP in 1998 as the standard codec for GSM; includes GSM-EFR and IS136 EFR (the TIA variant).

AMR-WB: AMR Wide Band, 50–7000Hz, from 6.60 to 23.05 Kbit/s, designed for UMTS.

Obviously, the latter codecs are suitable for VoIP in wireless NGN and IMS.

3.6.5 NGN Signaling

If the IETF deserves the title of the founder of the NGN, ITU-T deserves the title of pioneer.

In the 1990s, ITU-T launched the first architecture for audio and video transmission over an IP network, as H.323. They were the first to adopt RTP (IETF RFC 3550) in the media and border gateways layer. The main signaling protocol was H.225.0, based on ISDN access protocol.

H.323 was well suited for enterprise telephony — most VoIP PBXs supported H.323 well before SIP.

But around 2000, ITU-T failed to get sufficient vendor and operator attention for BICC, the signaling protocol allowing to set up media streams.

Conveying the media streams themselves requires routing technology, something the traditional telephone exchange vendors did not master in those days. Moreover, their energy was dragged away by the implementation of IETF's SIG-TRAN (Signaling TRANsport), to carry signaling over IP networks. It became tempting to phase out BICC until after ISUP over SIGTRAN.

The need to carry signaling over IP was so high that individual companies had developed Reliable User Datagram Protocol (before IETF SCTP) and Extended ISUP over IP (E-ISUP, before carrying ISUP over IETF M3UA).

The lack of adoption of ITU-T BICC opened a window of opportunity for the IETF to initiate the Session Initiation Protocol (RFC 2543) in 1999 and clarify it as RFC 3261 in 2002.

---

* Bit Error Rate.

IETF SIP-T (RFC 3372) is a binary encapsulation of ISUP inside SIP (as a MIME attachment), with the purpose of not losing information when transferring calls over SIP networks. 3GPP defined this as TS 29.163.

But as ITU-T ISUP was heavily modified and extended by national PTTs, which lead to over 60 national or regional variants across the world, SIP-T is also national.

A new variant, SIP-I is therefore currently under development as an international superset of the national SIP-T variants.

The ITU-T Study Group 11 is defining a new encapsulation called Generic Transparency Descriptor (Q.GTD.1).

The reason for these encapsulations is that even in 2006, it is not possible for an MGC* to map the full set of ISUP messages and parameters to basic SIP messages and fields, despite all additional SIP RFCs trying to cover the most evident holes (e.g., there's still no originally called number, redirecting number nor privacy indications in SIP).

Anno 2006, ITU-T ISUP, and BICC contain a superset of IETF SIP functionality, so the battle between *bellheads* and the *bitheads* is still raging.

3.6.6　NGN Gateway Control

Also in the NGN area, in 2000, the ITU-T initiated another major and ambitious development: an open interface between the session control layer and (initially only) the media gateways: Media Gateway Control (MEGACO, then recommendation H.248, today H.248.1).

Today's version (H.248.1 version 3, 2005) includes 13 basic packages and has over 40 optional packages to control various telecom features on circuit-to-IP media gateways and IP-to-IP border gateways. TISPAN defines Core Border Gateway Functions (C-BGF, at the edge between access and core network) and Interconnection Border Gateway Functions (I-BGF, interconnecting the core networks). The general term "gateway" could be used to refer to both media and border gateways.

ITU-T H.248.1 is adopted both by TISPAN (for "fixed" NGNs) and by 3GPP (for the "mobile" IMS).

The physical decomposition of session plane and user plane is now even adopted in the new generation of GSM (Phase 2+) and UMTS core networks (3GPP Release 1999 and Release 4), where centralized MSC servers are now controlling decentralized media gateways. Hence ITU-T H.248 has also become one of the reasons why IMS (3GPP Releases 5-6-7) is losing ground.

The IETF first adopted H.248 (as Megaco 1.0, RFC 3015) in 2000, but continued to develop its own Media Gateway Control Protocol (MGCP, RFC 2705 superseded by RFC 3435 in 2003).

Today we see MGCP losing ground against H.248.1, which is considered more "open" and "extendible."

---

* Media Gateway Controller, please refer to Section 3.1.

3.6.7 *Transmission*

ITU-T played a prominent role in the worldwide standardization of digital subscriber lines; the G series also contains the standards for ADSL and its higher speed flavors.

Also in the transmission layer, G.703 contains the physical/electrical characteristics of any digital hierarchical interface (e.g., the DS1 system: E1 at 2 Mbit/s, or T1 at 1.5 Mbit/s).

G707/708 specifies the Synchronous Digital Hierarchy (SDH) for transmission over optical networks (e.g., STM-1 at 155 Mbit/s, STM-16 at 2.5 Gbit/s up to STM-1024 at 159 Gbit/s).

3.7 IEEE

As mentioned in Section 1.7, the IEEE today plays a capital role in driving new wireless access technologies in both unlicensed and licensed frequency bands (i.e., WiFi and WiMAX, respectively), to carry IP.

In the history of telecommunications, IEEE 802.1-2004 ("WiMAX") and 802.16e-2005 ("Mobile WiMAX"), with the latter written as a delta specification on the former, are major milestones. At the start of the twenty-first century, these 800-page documents truly reflect the latest advances in radio technology, corresponding to the OSI* model's Layers 1 (PHYsical) and Layer 2 (Medium Access Control).

One could object that these are connectivity technologies rather than enabling communication services.

But the WiMAX standards introduce a revolutionary and long-awaited Meshed network architecture, where WiMAX Subscriber Stations (SS) are allowed to transmit to each other, obviously not at the same time on the same frequency. Though there are currently also examples with WiFi Meshing, 3GPP and 3GPP2 have held tight to the Point-to-MultiPoint (PMP) architecture, where one Base Station (BS) transmits to multiple Subscriber Stations (SS), but the SS don't transmit to each other.

So the IEEE WiMAX standards are worth a few words in this chapter on NGN standards, with the humble ambition to provide a very basic overview only.

Seen from the Network Layer 3, the WiMAX MAC Layer 2, by including the appropriate "Convergence Sublayer" (CS: the top part of the MAC) can behave as one of the following:

> An ATM Virtual Circuit (VC)
> An Ethernet (IEEE 802.3) Local Area Network (LAN)
> A Virtual LAN (IEEE 802.1Q-1998)
> An IP router (RFC 791, 2460)

---

* Open Systems Interconnection.

At the PHY layer, WiMAX is in fact a whole family of Broadband Wireless Access (BWA) technologies:

■ WirelessMAN-SC™ ("Single Carrier"; Wireless MAN SCa™ single carrier advanced): in the 10 to 66 GHz frequency band, wavelength is short (3 cm to 4.5 mm), hence multipath reflections against buildings and the landscape are negligible, so Line-Of-Sight transmission (LOS) is required. The transmitter and the receiver need "see" each other. With a channel bandwidth of 25 MHz, raw data rates in excess of 120 Mbit/s can be obtained.

   In the frequency band up to 11 GHz, reflections are significant, Non-Line-Of-Sight (NLOS) or near-line-of-sight operation is possible, but interference appears (i.e., receiving two signals with the same frequency from two transmitters). So either the available spectrum needs to be allocated in slices to individual operators (licensed mode), or the power at which these operators transmit needs to be limited (unlicensed mode, primarily in the 5–6 GHz band). In the latter case also, specific collision avoidance mechanisms such as Dynamic frequency Selection (DFS) may be imposed by the national regulator. Below 11 GHz, either in licensed or in unlicensed mode, there are three options for the PHY layer:

■ WirelessMAN-OFDM™ ("Orthogonal Frequency Division Multiplexing"): OFDM is a transmission technology where, using a channel bandwidth BW (e.g., of 5MHz) centered around a carrier at frequency f_c (e.g., 2302.5 MHz), 200 orthogonal subcarriers (e.g., spaced 22266 Hz from each other) are used to transmit a set of 200 complex numbers c_k during one Symbol time T_b (1/22266 Hz = 45 μs in this case). If, for example, c_{13} is zero, it means that bit 13 (out of 200) is zero. The 200 bits transmitted during T_b are visually represented as a QAM* constellation (dots on a grid, either lit or dark). Besides these 200 frequencies, as 256 is the smallest power of 2 above 200, 56 frequencies are used as a "guard band" transmitting zero Voltage, in order not to interfere with the next channel (e.g., 2305 to 2310 MHz).

   The transmitter uses an Inverse Fourier Transform (IFT) from frequency to time domain, in order to derive the voltage being sent to the amplifier and the antenna.

   When the receiver applies a Fast Fourier Transform on the signal, it recalculates these 200 complex numbers c_k, and if they haven't moved too much, meaning if the QAM constellation dots were sufficiently apart, discovers the correct value of the 200 bits.

   The bandwidth efficiency in this simple but realistic example is 200 bits × 22266 Hz = 4.453 Mbit/s, over 5 MHz bandwidth, thus almost one bit/s

---

* Quadrature Amplitude Modulation.

per Hertz. In real life, that value is a bit higher for downlink and lower for uplink traffic, but never mind, is roughly double of what you can get out of CDMA transmission (3GPP HSPA and 3GPP2 EVDO).

- WirelessMAN-OFDMA™ ("Orthogonal Frequency Division Multiple Access"): Here, the subcarriers (e.g., 10, adjacent or not) are grouped in subsets (i.e., 20) named subchannels. In the uplink, several transmitters may transmit simultaneously as long as they use different subchannels (e.g., 3) or other times (e.g., 4 OFDMA symbols) to transmit — hence the name "Multiple Access." The combination of a subchannel and time is called a slot; in the example there are 12 slots available.
- In unlicensed mode below 11 GHz, WirelessHUMAN™ is the term IEEE uses to define the usage of one of the three PHY options above, plus the specific requirements for unlicensed operation.

It is WirelessMAN-OFDM (licensed or not), and thus also WirelessHUMAN (unlicensed), which support both PMP and Meshed network architectures.

As for any two-way radio technology, WiMAX uplink and downlink transmission either have to use different times (TDD: Time Division Duplex) or different frequencies (FDD: Frequency Division Duplex); both modes are supported in WiMAX. GSM is only TDD and for 3G, incompatible variants have appeared:

> Chinese TD-SCDMA is obviously TDD.
> 3GPP W-CDMA is FDD.
> 3GPP2 EVDO is FDD.

Besides WiMAX, and contrary to 3GPP/3GPP2, which only focuses on long range wireless technologies, the IEEE also developed a low-power, ultra-short range (10m) wireless technology, Bluetooth (802.15.1), to build Wireless Personal Access Networks (WPAN) or wireless peer-to-peer networks.

The short-range wireless technologies benefit from rapid and worldwide adoption by the computer and peripheral manufacturers.

With the mobile cellular phone now a computer, IEEE technologies find their way into these devices, perhaps at an even faster pace than 3GPP/3GPP2 technologies.

3.8 OMA

The OMA was created in June 2002 in response to various organizations attempting to standardize mobile data services: the WAP Forum (focused on browsing and device provisioning protocols), the Wireless Village (dealing with instant messaging and presence), the SyncML Consortium (for data synchronization), the Location Interoperability Forum, the Mobile Games Interoperability Forum and the Mobile Wireless Internet Forum. The OMA was created to gather these initiatives under a single umbrella.

OMA is seriously addressing messaging services with the Multimedia Messaging Service (MMS) and Instant Messaging and Presence Service (IMPS), today widely implemented on mobile phones, in frontal competition with IETF's XMPP, with Jabber and of course IM provider-specific protocols.

OMA Data Synchronization and Device Management are defined using the syncML (synchronization Markup Language).

Further, there's a standard for mobile broadcasting (BCAST).

OMA is today the best promoter of GSM/GPRS/UMTS mobile data services and matching handsets, but will this model prevail over the TISPAN NGN or — even more difficult — the wide open Internet?

3.9 Summary

In this chapter we started from a layered definition of the NGN, and after having reviewed the work of the most relevant standardization organizations one by one, we can now draw their areas of focus on that initial diagram.

In conclusion, though the standards organizations are often endorsing each other's specifications, they seem to have found their place in the NGN.

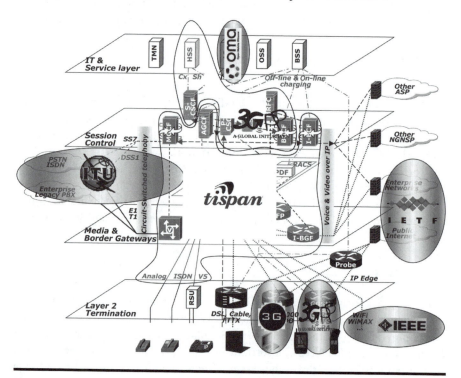

Figure 3.6   Standards bodies and their domains of contribution to the layered NGN.

Some areas get little attention, such as Media Resource Function Controllers (MRFC), Media Resource Function Points (MRFP) and Media Gateways — especially when migrating from voice to video or new media types.

Another important no-man's-land is the IT part (OSS, BSS and TMN); whereas today, 40% of the effort to deploy a new service probably goes into provisioning, billing and supervision tools and processes. Organizations that used to cover this area, for example, the Object Management Group (OMG) with a.o. the Common Object Request Broker Architecture (CORBA), have lost ground and momentum. Rather than focusing on the telecom sector, they tried to be everything to everyone, including the financial sector.

Fortunately, the times where a different radio access technology (GSM, CDMA) would lead to different core network architectures and even different VAS platforms, seem to be behind us. As a result of this bitter fight, 3GPP and 3GPP2 lost a lot of authority and credibility. GSM has won that battle, with 85% of subscribers worldwide.

This has left an open space to TISPAN, to define the PSTN Emulation Subsystem (PES) and make advances in network attachment, resource and admission control and service-based policy decisions, based on IETF (DIAMETER) and ITU (H.248) protocols.

However perhaps on a larger scale, IETF and ITU are now competing to adopt and globalize the TISPAN standards. TISPAN seems to push toward ITU.

The battle between bitheads and bellheads is likely to be pursued for the foreseeable future.

And at an even greater scale, communication standards will keep competing with powerful vendor-specific solutions at all layers of the NGN: peer-to-peer communication technologies, regional HD TV formats, closed instant messengers and search engines, incompatible music and video formats, proprietary radio technologies such as Flash-OFDM or WiBree.

Chapter 4

Engineering the NGN

Up to here, we have examined the fundamentals of genuine communication services (Chapter 1), considered at a modern service provider's starting position (Chapter 2), and taken a tour through the landscape of NGN standards (Chapter 3).

What does it take to become a NGN Service Provider (NGNSP)?

As we'll see in the upcoming chapters, we'll need a way to charge our customers (Chapter 5) and a reliable VAS platform (Chapter 6).

But here, let's concentrate on the practical way to build a NGNSP, which should lead us to consider some essential business aspects too.

To start quickly, we could download the SIP Express Router (SER) and SIP Express Media Server (SEMS) from iptel.org on a Linux PC, and be ready in minutes to convey voice and video traffic over IP. Let's call our little enterprise ngvas.com.

The next step is to persuade Alice White,* our model user, to configure her IP phone, softphone or multiplayer game to use ngvas.com as the default proxy for outbound voice and video communications.

Would Alice submit her outgoing session initiation requests for sip:19191234567@ngvas.com, sip:19191234567@chicago.com or even sip:carl.sandburg@chicago.com† to ngnvas.com?

How would Alice know of the existence of ngnvas.com — through search engines? Would her IT administrator in atlanta.com look for new VoIP carrier

---

* Alice is the originating user in IETF RFC 3261 on SIP. Alice White was an American film actress born as Alva White (1904–1983), who started her career in silent films but successfully switched to talking movies in the late 1920s, as her voice matched her screen image perfectly.

† Carl Sandburg (1878–1967) was an American poet, historian and novelist who focused most of his poetry on Chicago, praising American agriculture and industry, a. o. in his famous poem "Chicago" in 1916.

Figure 4.1   Alice White, actress.

reviews? Would her IT administrator even download a local softswitch from ngnvas. com, or rent some capacity through some Web Service API to ngvas.com?

More fundamentally, when peer-to-peer instant messaging, voice and video over the Internet are free, and bandwidth is abundant, what will drive Alice (e.g., end users, companies, third-party software) to use centrally operated communication services in the client–server model?

Should ngvas.com strive to become a SIP registrar and SIMPLE presence server, allowing third parties to query Alice's presence for inbound communications, rather than establishing outbound communications?

To whom would Alice grant the permission to reveal the status information regarding the address sip:alice.white@atlanta.com? Would that include her location information? Would her existing Outlook calendar be synchronized with ngvas.com?

Would ngvas.com be able to offer a better outbound voice and video quality than its competitors (centrally operated or peer-to-peer)? And if so, would calling users accept to pay a premium for it?

Figure 4.2   Carl Sandburg, poet, historian and novelist.

And if Alice is convinced to purchase outbound communication services, or to publish her presence for inbound communications, what technical hurdles will she or her company IT administrator face?

It is these kinds of questions we need to tackle in this chapter; hopefully this example will lead us to new insights in essential network engineering requirements.

Ten years ago ngvas.com could not have appeared due to the entrance fee to purchase and operate the technical infrastructure.

Today, there's affordable technology and some solutions for 24x7 operations (e.g., hosting), but ngvas.com would most likely fail due to the complexity of Internet business models and commercial chains.

Up to now, network engineering has been about topography, redundancy, predicting the traffic to be carried and dimensioning the infrastructure accordingly.

Today, as the distinction between a network and a service is almost gone, "Engineering the NGN" is to be interpreted in a totally different way.

Nevertheless, in this book we still think it's worth maintaining the distinction between "NGN" and "VAS" even if they might be offered under the same commercial label (ngvas.com).

4.1   Advertising and Discovery

One of the reasons why communication consumers used to rely on central communication service providers is the addressing plan. We already mentioned this as a strength in Section 2.1. As the service providers are mutually interconnected in a cascade model, end users are able to communicate with a large number of easily addressable destination users, identified by unique and permanent contact addresses, that is, E.164 telephone numbers.

Communication consumers have a fundamental need, namely to resolve these higher level contact addresses (URI, Uniform Resource Identifiers, RFC 2396) to discover a routable IP address — either directly of the destination end user, or more likely of a suitable border gateway, able to relay the media streams onwards to the actual destination network or user.

Whereas for classic telephony the originating user would rely heavily on the communication service provider as an "address translator," the ubiquitous availability of always-on IP connectivity (broadband) and the packetization of our communication means are changing that model. The end user today has a wide range of options to discover the routable IP address corresponding to a buddy in a contact list or (maybe soon) a geographic location on a map, a character in an online game, and so on.

Domain Name Service queries (DNS, RFC 1035 and related) may be performed to obtain host names and finally routable IP addresses for various application addresses, for genuine communications or for the transfer of stored media, for SIP- and non-SIP based applications. A DNS server can, for a single destination domain, hold different records* for SIP and for other applications (e.g., e-mail, browsing).

The public DNS infrastructure has not been populated with records for SIP, but if they wanted, ISPs could thus use that general public mechanism to direct SIP requests from their broadband users to the NGNSP's.

Alternatively, the ISP could allow each broadband customer to select a default NGNSP, or even to indicate their preferred NGNSP in the SIP URI, on a per session basis (sip:carl.sandburg@chicago.com.ngvas.com).

In an even more advanced scheme, the user agent may choose to send all SIP requests to an address resolution service (locally preconfigured as an IP address or

---

* IETF RFC 3403 defines these as the Naming Authority Pointer (NAPTR) Resource Records (RR).

a URI'*), which would respond with SIP 301 "Moved Permanently," returning a Contact:[†] leading to the actual NGNSP[‡].

That SIP URI resolution service could be operated independently, or be part of a real NGNSP (i.e., carrying media too, according to our definition in Section 3.1).

It may even be developed as a peer-to-peer SIP (P2PSIP) service, ultimately redirecting the user agent's request to the physical server responsible for the requested URI.

Multiple technical options are thus available for the selection of a NGNSP.

Whatever option is taken by user agents and people distributing them (we'll talk about distribution in the next section), the advertising and discovery of NGN networks boils down to a quite simple question: will existing name resolution algorithms and processes (e.g., DNS, SIP) be able to discover the *best* NGNSP for a given destination address?

Early attempts to answer this question had resulted in TRIP (Telephony Routing over IP, RFC 3219, 2002), a routing protocol to advertise the reachability of telephony destinations (E.164 numbers and ranges), independently of the application (signaling and media) ultimately used to reach those destinations. By listening to TRIP updates from neighbors, a NGN network element can construct a Telephony Routing Information Base (TRIB): a database of both internal media paths within the IP Telephony Administrative Domain (ITAD) and external media paths to different ITADs.

DNS is of course the general name resolution protocol used by most Internet applications to reach a destination address. Therefore, the IETF also undertook the tElephone NUmber Mapping effort (ENUM, RFC 3761, 2004) recommending the use of DNS to obtain Naming Authority Pointer Resource (NAPTR) records, containing the various URIs (e.g., Universal Resource Information for SIP, e-mail, classic telephony, instant messaging) in order to reach a given destination telephone number (in the format 8.7.6.5.4.3.2.1.8.7.9.1.e164.arpa.).

However, whether we use TRIP or ENUM to resolve destination NGN addresses, it will always result in media streams (RTP packets) being routed directly to the final NGNSP (recipient operator) "owning" the destination user.

Indeed, the destination-based forwarding paradigm results in the most specific route being used. If an intermediate NGN provider has advertised to be capable to route the entire number range +1978XXXXXXX, but the final NGN provider advertises to be responsible for the range +1978123XXXX, both TRIP and ENUM will result in the final NGN provider being selected.

RTP packets will thus be routed end-to-end, from the originating NGN provider's border gateway to the public IPv4 address of the final NGN provider's border gateway, along a route discovered by IP routing protocols.

---

* For example, sipuriresolutionservice.com, to be resolved by public DNS.

† SIP/2.0 301 Moved Permanently. Contact: "Bob" <sip:bob@biloxi.com.sip.thebestngnserviceprovider.com>

‡ From thebestngnserviceprovider.com.

There are two categories of IP routing protocols: Interior Gateway Protocols (IGP) to be used within an administrative domain (AS: "Autonomous System"; e.g., an ISP), and Exterior Gateway Protocol (EGP) for use on inter-AS interfaces.

Within the IGP, distance-vector routing protocols (e.g., Routing Information Protocol, RIPv2, RFC 2453) advertise routing information as the distance to a destination IP address range, but each router is unaware of the network topology. The propagation of new routing information can be quite slow. With link-state routing protocols (e.g., Open Shortest Path First, OSPFv2, RFC 2328), each router has the same link-state database describing the entire network topology of the AS: "Autonomous System," administrative domain). Link-State Advertisements (LSA), when broadcast throughout the AS, can form quite some overhead traffic. Therefore, both distance-vector and link-state routing protocols are not suitable on inter-AS interfaces.

The most commonly used EGP is the Border Gateway Protocol (BGP-4, originally RFC 1105, now RFC 4271, 2006), suitable for use on inter-AS interfaces.

TRIP is based on, and very similar to BGP-4. The Autonomous Systems (AS) of BGP-4 correspond to IP Telephony Administrative Domains (ITAD) in TRIP. Where BGP-4 lets IP routers advertise NEXT_HOPs, TRIP lets location servers (LS) advertise NextHopServers.

BGP-4 and TRIP share the same fundamental limitation: the inability for the originating service provider to select a transit AS resp. ITAD in order to reach a destination AS/ITAD, *if* the originating service provider has received routing information concerning the destination AS/ITAD. This is of course often the case.

In summary, today it's fair to state that NGNSPs are blind for media quality metrics, and are unable to select transit NGNSPs objectively offering better media quality to reach a given destination. The existing query/response protocols (SIP, DNS and ENUM) or telephony routing protocol (TRIP) could be used to communicate better next hops, but the oversight is missing. The result is that media compete for bandwidth with less critical applications on the same IP routes. The situation is comparable to what aviation must have been before the birth of Air Traffic Control.

There appears thus to be a need for a new intelligent overlay, that would be aware of the media resources, throughput and quality in the media and border gateways layer, and able to return better next hops when queried by the session control layer.

On a busy Friday afternoon, the media stream from Alice in Atlanta to Carl in Chicago, and the backwards stream, should perhaps be routed like this:

Could ngvas.com also be launched as such "advertising overlay," not conveying the actual media streams, and not even the session establishment signaling?

It looks like an attractive perspective, but either ngvas.com would have to build its own network of probes, permanently making test calls using various providers to various destination domains; or, more realistically, there needs to be a mechanism

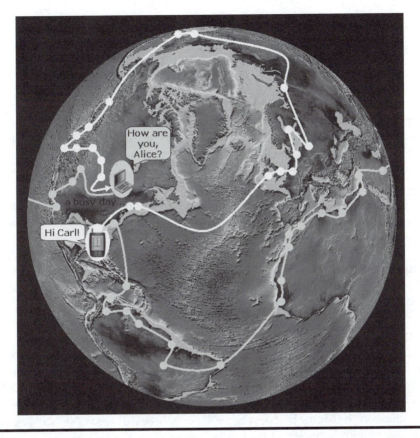

Figure 4.3   Media streams avoiding a congested zone of the Internet.

to obtain objective information from the existing NGNSPs, who in turn are relying on IP service providers — the transport layer.

So in the next section, let's start from what should be the first concerns of any NGN communication service provider: how to provide media throughput and quality over third party IP networks and the public Internet?

4.2   The Quest for Throughput and Quality

Much has been said about quality of service (QoS) and, as for security, these topics are too large to be treated from a single angle.

Also, it means something else to different parties. ISPs would think in terms of resource reservations along an IP route, prioritizing specific IP packets within each router, labeling packets using MPLS, providing virtual private networks, and so forth. The 3GPP IMS, another example, promises "end-to-end" QoS enforcement (as we reviewed in Section 3.3.2), but assumes an IMS at either end of the communication and no intermediate service providers. End users would probably think of "uptime" when talking about quality of service.

Let's go through a list of questions to figure out what NGNSP's should target when talking about quality.

Will NGNSP's take a defensive attitude, and let their softswitches interrogate some local TISPAN Resource and Admission Control Subsystem (the RACS we discussed in Section 3.5) to verify whether there's enough bandwidth to establish the session the customer wants?

Would that be based on the proposed media type (offered SDP) or, better, on the negotiated one?

Could the NGNSP then, after SDP negotiation, simply refuse to establish the session — what release cause should be invented for this and will the end customers accept this?

Or should the NGNSP, over a given IP access bandwidth, try and demonstrate better end-to-end media quality (voice and video) than ISPs currently calculating the best IP route, by calculating the best media path? That would be the pro-active strategy.

Who can rightfully (both commercially and technically) claim to know the better next hop for media and other applications?

Is it the same best next hop for all applications, or could it be the NGNSP controlling the best media resources toward a list of specific destination domains, as measured via RTCP XR? (Real-Time Control Protocol's eXtended Reports, IETF RFC 3611, briefly mentioned in Section 3.2.2)

Our hope is that the best next hop will not only be a discount NGNSP offering the lowest price regardless of quality, or a super virtual private network with the best quality at an unaffordable price. There will be a wide palette of opportunities between the extremes, if there's an objective and independent way to assess a quality increase above best effort.

Will specialization occur, with some NGN providers guaranteeing bandwidth (for video), vs. others offering the lowest round trip delay and packet loss rate (for voice)?

These are the questions we would need to investigate before positioning and launching ngvas.com, our little NGNSP.

Intuitively, one can imagine that by diverting media (and real-time applications in general) to less occupied IP routes than the ones calculated by the ISPs, some aspects of media quality could be improved.

Peer-to-peer applications, which according to ISPs today account for 50 to 75% of all traffic on the Internet, have demonstrated that by splitting a single media stream over multiple peer relays, and recombining the streams at the receiving end, better throughput and quality could be obtained. Besides for genuine communication services, peer-to-peer networks are also better at distributing stored content (file sharing). In fact, peer-to-peer networks are currently "eating" the bandwidth originally deployed to be used by client-server applications such as browsing or email.

Can the NGN, essentially using client-server protocols and not splitting up media streams, fight the peer-to-peer phenomenon?

Not today, that's our impression, not in the present state of the standards.

But better times might be ahead.

We introduced H.248 in Section 3.6.6. In 2004, ITU-T standardized an H.248 package allowing the media and border gateways to communicate the RTCP XR measurements to a higher layer: the ITU H.248.30 package. With this package, objective measurements can be collected either per segment, or for an end-to-end media path, but also subjective metrics: the estimated Mean Opinion Scores (MOS) which end users would have granted, for Listening Quality (MOSLQ) and Conversational Quality (MOSCQ).

Other H.248 packages allow the gateways to report on congestion (H.248.10, H.248.32), overload (H.248.11) or jitter buffer metrics (H.248.31).

RTCP XR measurements could eventually be obtained from a network of independent test call emulators using H.248.30 to report, but then the other H.248 packages would not be covered.

Similar metrics can be used to assess video quality.

Selected friendly users could also be invited to report on quality, during or after each call, and be rewarded for this.

Imagine that it would be the Meshed Mobile WiMAX* Subscriber Stations (SS) doing the RTCP XR measurements — the metrics would constantly change as the SS would move around. It would become a computing-intensive job, that's for sure.

Communication buyers — initially the NGNSPs themselves, maybe later the end users — are thus going to be looking for independent organizations receiving this information, exchanging it peer-to-peer, analyzing it in detail and delivering independent, objective routing advice, as fast as possible.

Within such framework, the "advertising overlay," NGNSPs should also be able to communicate on other aspects such as price, available total capacity, network roll-out plans or current number of customers.

The idea is thus a "stock exchange" for NGN communication services.

The common goal should be to beat peer-to-peer media conveyers, not by barring their port numbers in firewalls or sabotaging their traffic in other ways, but through an objectively and subjectively better service.

A tough battle for ngvas.com, our little NGNSP.

The idea for networks to advertise their quality through standard protocols is not new. It was originally the case in GSM international roaming, for example, where the visited networks would announce the minimum radio signal level (RxLevMin) a mobile would need to receive from that network, in order to have a decent quality voice conversation. If multiple networks would be available, the mobile handset would select the network it received the best, if that network had

---

* The WirelessMAN-OFDM™ with IP Convergence Sublayer – please refer to Section 3.7.

not announced a too high RxLevMin compared to the actually measured reception level. At similar reception levels, the handset would opt for the network announcing the lowest RxLevMin.

The first thing that happened is that all operators started putting RxLevMin at the minimum value in order to capture inbound roamers, even when call quality would not be OK. Who would these visiting roamers call to complain anyway?

The network selection system was later further torpedoed using Over-The-Air (OTA) provisioning, over SMS, to populate the Preferred Networks list in your SIM card.

Today when you go roaming out abroad, the selected network is rarely the fruit of hazard or objective quality considerations. Either the visited operator will be chosen within the same multinational operator group, or in other cases there's a bilateral commercial preference agreement beyond the standard roaming agreement.

The lesson to be learned is that in the long term, communication service providers cannot be trusted to announce objective network quality.

Probably similarly, public companies could not be trusted to announce correct financial reports to the stock exchange.

Even when it would be operated independently from start, the "communication means advertising overlay" is thus likely to evolve from resembling a Stock Exchange, to a Securities and Exchange Commission.

4.3 Distribution

Once advertised (perhaps through an independent organization and peer-to-peer network) and ensured (in terms of throughput and quality), a service must be distributed.

Today's IM, voice and video providers are using a simple business model: make your user agent free for downloads, use a proprietary protocol and addressing scheme, build up a captive community and let your advertisers pay for the bandwidth being consumed by your central servers. "Don't worry about quality; your end users are captive." Sure?

The business model of an NGNSP, using standards, will be *very* different.

User agents will be downloaded for free and most likely not be controlled by the NGN providers at all; remember Section 1.6. Most likely, user agents will also built into other applications such as navigation, gaming or in-car entertainment. Device manufacturers will continue worrying about the forces of the global consumer electronics market, and be unable to predict whether the SIP user agents will be installed on video camera's, in cell phones or in home theatres. It's not a matter of unwillingness.

The distribution of basic NGN service could thus become a thin, subtle process. Even existing and well-established IM, voice or video providers could add the standards to their existing software clients, and let consumers choose between peer-to-peer, non standard and standard client-server mode of operation. Given the quality metrics we discussed in Section 4.2, the client could take its own decisions, or rely on an independent third party.

There will probably also be no relationship possible between the connectivity service provider and the communication service provider; the idea to convince ISPs to direct all SIP requests from their broadband users to a selected NGNSP is flawed. As there's no regulator imposing to implement some kind of carrier selection, any agreement would have to be commercially negotiated. But with competition from free peer-to-peer networks, could NGNSP's pay the ISPs enough to do so? Commercially, can the ISP ask or even impose its customers to select a NGNSP?

The situation is similar to GSM service distribution in advanced markets. The GSM service provider in most cases still provides a radio network, a core network and some basic services such as voicemail or roaming — even though the engineering and daily operations are today being outsourced to the vendors of these networks.

The commercial distribution is moving from a retail model to wholesale. It is outsourced to a very capillary network of franchise holders and distribution chains, even increasingly using their own brand. There's a semantic debate on when such reseller becomes a "Mobile Virtual Network Operator"; some say it's enough to offer your own self-care Website, others claim you should be operating own billing system and prepaid VAS platform, or even distributing your own Subscriber Identity Modules (SIM). For sure, resellers have been able to define their own tariff plans since the launch of GSM in the early 1990s.

A similar commercial distribution model is to be expected for basic NGN service; perhaps the distinction between retail and wholesale will disappear. Anyway it will be hard to tell the difference.

As for mobile cell phone service, the demand for NGN communication services will most likely start in the upper enterprise segments — a retail model very close to wholesale.

Open source PBX's, hosted PBX services and desktop conferencing tools are today confined within a single enterprise or even enterprise site. When these islands of IP will start communicating, that is, along the customer-vendor supply chain, there will be the need for protocol and media conversions, for address analysis and routing, for network-based message stores, and so on — for NGN/VAS service providers. The TDM-to-IP migrations in the enterprise segment are also laying the foundation for tomorrow's demand for large-screen video and hi-fi quality audio (telepresence).

As for cell phones, NGN technology and the demand for media quality will spread from the board room to the work floor, from enterprise to residential usage, from VoIP telephones and PCs to the other devices, and finally from the online world to off-line content downloads.

In 2006, the first digital media kiosks have appeared at airports and in stores.

Cable operators are demonstrating loyalty to the IETF and TISPAN NGN standards, at least in the telephony part of their triple-play offers.

The media industry, and more specifically broadcasting, is heading for a major clash with the telcos and ISPs if it doesn't embrace these standards in the early digital TV and radio deployments.

Before selecting a retail or wholesale strategy for ngvas.com, or even focusing on the corporate enterprise segment, let's therefore consider an important aspect of the offer: is ngvas.com going to offer outbound or inbound communications?

4.4 Outbound Communications

Today's public networks and the Internet allow reaching almost any destination number or address on the planet.

This has been realized through almost a century of carrier interconnections, and twenty years of Internet service provider peering.

Conveying these communications over IP, using an NGN infrastructure and possibly an advertising overlay to improve the quality and availability, may sound as a good plan toward countries well-connected to the Internet. But there will always be these hard-to-reach Internet destinations, which could easily deteriorate the perceived overall quality offered by the NGN provider, certainly in the business and corporate communication markets.

Media gateway operators, providing the packet-to-circuit conversion toward these countries (in fact, today's international voice carriers), should of course also be able to advertise their termination services on the peer-to-peer overlay. They should then provide the same media quality metrics and advice-of-charge information as the "IP-only" NGN providers.

In fact, many multinational enterprises are today using such international voice carrier, even for internal calls, which could be carried over the enterprise's "own" IP backbone. Enterprise PBX users are today forced to dial 9 or 0 to get an outside line, often to the local media gateway, and as soon as they do, their PBX is incapable of recognizing that the destination is internal (but on the other side of the world).

By switching from a "carrying" to an "advisory overlay" role, the international carriers could start using these corporate IP backbones to carry media streams. On circuit-switching PBX sites, the media gateways converting circuits to packets could be operated by the carrier. The sites where IP telephony is introduced (i.e., where a commercial or open source IP PBX is deployed to replace the legacy PBX) can be used as a transit points for signaling and media.

With the necessary security precautions, and under responsibility of a trusted NGNSP, the media streams of enterprise A could even make use of the softswitches and media relays of enterprise B, and vice versa.

Initially both enterprises A and B should probably be customers of a single NGNSP; later, NGN peering agreements will appear between competing NGNSPs.

One can see a scheme where not only service providers and carriers, but also enterprises are rewarded for their contribution to the NGN maze.

In order to realize that vision, we are facing the same limitation, namely, that today's IETF and TISPAN standards do not allow to properly advertise the availability, quality and price of media resources.

In conclusion of this little section, using enterprise IP networks to convey outbound communications sounds original and ngvas.com could specialize in turning enterprise telephony networks into NGN communication service providers; first for their internal use; later maybe as part of a larger federation.

4.5   Inbound Communications

Perhaps without noticing, we have looked at NGN advertising, media quality and distribution from the angle of outbound communications.

In the 1980s, International Voice Carriers (IVC) started their business by providing outbound international and long distance telephony service to enterprises. The incumbent PTT would continue to deliver the inbound calls to the E.164 numbers, that is, over ISDN Primary Rate Interfaces (PRI) to the PBX.

Throughout the 1990s, even through new access technologies appeared to bridge the last mile (e.g., microwave transmission), and the carriers were rolling out their metropolitan, national and international glass fiber loops, the interconnect revenues the alternative carriers could receive would not justify handling the inbound communications to the enterprises. Besides transmission, the cost of the digital telephony exchanges (switching) would also be a show stopper.

Competitive Local Exchange Carriers (CLEC) and Local Number Portability (LNP) were introduced by the US Telecommunications Act of 1996. But even then, CLECs would have to lease digital exchange capacity from the Incumbent LEC (ILEC) and thus would not be able to introduce lower-cost switching technology (NGN). Local Loop Unbundling (LLU) — where a CLEC only has to lease naked copper pairs from the ILEC — was only introduced in the early 2000s in advanced markets (2003 in the U.S.).

LLU and DSL standardization have permitted the emergence of affordable broadband IP connectivity, together with new access technologies such as Cable, WiFi and WiMAX.

Now residential users and businesses can enjoy affordable broadband IP access, and with the emergence of VoIP technology and devices, new providers are taking a fresh look at inbound communication handling. The telephony interconnect revenue is low per minute, and the cost of interconnection is not negligible, but the available broadband capacity for VoIP termination is high. The added value is high too: call centers can now be centralized and located anywhere, taking inbound calls collected on local E.164 numbers in dozens of countries. And the opportunity to provide presence-based VAS is looming — the basic force we discussed in Section 1.8.

Hence a new type of communication service provider has appeared in the mid-2000s: the Inbound Communications Handler (ICH), operating TDM-to-VoIP media gateways, or leasing them from the International Voice Carriers (IVC, who were already using them in the other direction anyway).

ICH can either apply for their own slice of E.164 numbers, comparable to CLECs, or start porting in local numbers from the ILECs or CLECs.

By the end of the first decade of the twenty-first century, maybe under pressure of unlicensed broadband mobile connectivity (Mobile WiMAX, IEEE 802.16e-2005), we can also expect licensed broadband mobile connectivity (3GPP HSPA and 3GPP2 EVDO) to become more reliable and affordable — including when roaming abroad. ICH will then be able to terminate voice and video calls to mobile devices — a privilege today reserved to licensed mobile operators.

Will 3G-only connectivity service providers end up teaming with the ICH, in their battle for market share against the mixed 2G-3G service providers and their 2G amortized networks? Or will the future Mobile WiMAX ISPs come to an agreement with the ICH? The general trend is to split connectivity service providers from communication service providers, but equal-to-equal partnerships will be possible.

In markets where the called mobile user pays for fixed-to-mobile calls (mainly in the U.S.), ICH should see a healthy stream of revenue straight from the called user.

Elsewhere, where the calling party pays, revenue will come from the interconnection. In fact, in such market, an ICH could attempt to reward the receiver of inbound communications (individual or enterprise) with part of the interconnect revenue. ILEC and CLEC don't do this today; on the contrary, enterprises need to pay monthly ISDN PRI rental fees to receive inbound calls. Some mobile service providers do reward their users for receiving calls from other networks. The NGN ICH play looks quite good; could our little ngvas.com be started as one?

The ILEC has the regulatory obligation to terminate calls to any national geographic number range assigned to, or numbers ported to the CLEC, and in this case to the specialized ICH. And the ILEC of course pays the call drop-off charges (call set-up fee and price per minute) to the CLEC/ICH, as specified by a national reference interconnection offer. Even the interconnection infrastructure to terminate these calls (leased lines) is, according to most regulation, fully at the expense of the ILEC; conversely, trunk groups being used for CLEC-to-ILEC calls are entirely at the charge of the CLEC.

By 2008–10, many regulators and ILEC are expected to introduce VoIP interconnections as a cost-effective alternative the E1/T1 leased lines — a loss of revenue for the ILEC's transmission wholesale department.

Worse, as a new cost, the ILEC will incur the costs of media gateways performing the circuit-to-VoIP conversion for a while, as long as many calls keep being originated in circuit-switched mode.

Regulators and service providers have some hard work ahead, extending today's circuit-switched Reference Interconnection Offers to the packetized world, thus transforming them into NGN peering agreements.

This brings us to the subject of our next section.

4.6   NGN Peering

TDM* (circuit-switched) communication networks have been *interconnected*. Bilateral interconnection agreements have been set up between the communication service providers (CSP). Without going into too much detail, these bilateral agreements are based on a connection fee per call, and a price per time unit. CSP X is charged by CSP Y to connect outbound calls (X to Y) to a given destination (e.g., Lebanon). And CSP X is also charged by CSP Y for inbound calls (Y to X) to free-phone numbers, or when CSP Y provides the physical access network allowing to select CSP Y as a carrier. Inter-operator accounting is based on call detail records, which are generated for every call. At the end of the month, the net amount due by CSP X to CSP Y (or vice versa) is calculated, in a process called "settlement."

Internet Service Providers (ISPs, "Autonomous Systems" in Internet standard terms) are however establishing *peering* agreements. Originally, ISPs would use settlement-free peering, reflecting the fact that traffic was symmetrical and both parties would automatically benefit from peering.

This is no longer the case. In a laconic recommendation D.50 approved in Montreal in 2000, the ITU-T spoke of commercial agreements taking into account the traffic flow[†], the number of reachable destinations[‡] and even the "geographical coverage"[§] as elements of the compensation, even when the actual transmission (the leased line) would already have been paid for 100% by the new ISP or country. In practice, it meant that small countries wishing to connect to a large public Internet eXchange Point (IXP, often in the United States) could be dismissed with a high offer, whereas most of that IP traffic (browsing, e-mail) would actually be routed to servers outside of the United States.

Contrary to TDM interconnection, the principles of ISP peering were thus left to be organized by each national regulator (for public peering — many little ISPs, low capacity), but most often by the private sector (private peering — high-capacity peering between the larger ISPs). Today, most of the Internet traffic occurs at private peering points, often implemented as third party "carrier hotels."

A starting pure NGNSP will thus have to go through multiple stages to obtain Internet access: first probably commercial broadband Internet access, then public peering, and then private peering over time — unless of course the NGNSP teams up from the start with a large ISP.

NGN peering is the interconnection of session-based networks. A session has a duration but is essentially set up to transport a certain volume of data (megabits). Will NGN peering then follow the principles of TDM interconnection (mainly time-based), or of ISP peering (volume, number of reachable destinations, etc.)?

---

* Time-Division Multiplexed.
† But what traffic flow: the estimated rate in Megabit/s or the actually measured rate? The average rate or the peak rate? Based on which formula? The principles out there are numerous.
‡ IP routes, reachable DNS domains, etc.
§ The number of square miles of the country?

It's a *very* difficult question. Today's NGN are mainly rolled out by incumbent fixed telephony providers (the ex-PTTs) for PSTN emulation (see Section 3.5) — they automatically think in terms of connection fees and duration. However, often they also have an ISP division, which thinks in terms of Megabit/s and destinations. And greenfield NGNSPs come and knock at the ISP door first. What will they then charge their peer NGNSP's?

If greenfield NGN providers do not start charging per minute, that is, because the general trend is to offer flat-fee, all-you-can-eat models, there will be no way for them to close TDM interconnection agreements to drop off outbound traffic to E.164 numbers or to purchase this traffic from international carriers. Greenfield NGN providers would thus have to focus on inbound calls for a while, for example, to mobile devices as we saw in Section 4.5.

Perhaps greenfield NGNSP's should simply not offer the E.164 telephone numbering scheme at all. Alice could only sip* Cal Sandburg at sip:carl.sandburg@chicago.com, eventually relayed by the biloxi.com domain.

Seen from the other angle, if the "ex-PSTN" NGN providers do not start peering based on volumes (and other parameters) instead of minutes, they will not be able to offer more advanced SIP-based end user services such as video-on-demand, and will continue to emulate the PSTN only.

There's a need for TDM interconnection and ISP peering agreements to converge to a new type of agreement, the ***NGN peering*** agreement.

In general, a new NGNSP needs to be the following:

> **Interconnected to existing TDM Core Networks** of fixed, mobile service providers and long-distance carriers, via Media Gateway Controllers (MGC) and Media Gateways (MGW†). The all-IP MGC (i.e., speaking ISUP over SIGTRAN) is connected to the TDM (SS7) networks via a Signaling GateWay (SGW). Service providers refer to the entire conversion of NGNs to TDM as "TDM trunking." In fact, as real-life NGNs are less monolithic than what the standardization organizations describe (Figure 3.1), trunking sometimes ends up being performed and operated by separate, dedicated organizations that have already deployed SGW and MGW in different countries, and are operating central MGC. Trunking is country-specific and relies on a very different skill set than deploying next generation networks: ITU-T or ANSI SS7 point codes, national SS7 ISUP flavors, SS7-to-SIGTRAN signaling gateways, transcoding, E1 or T1 leased lines, fixed operator licenses, interconnection agreements, in-country local premises or hosting, and so on. If the NGNSP decides to outsource TDM trunking, he will need to peer with his TDM trunking providers via IP (SIP and RTP).

---

* Yes, we'll have to start using this verb in a totally different context.
† These are referred to as Trunking Gateways (TGW) in the TISPAN standards.

Interconnected to existing TDM Access Networks, using TDM-to-NGN conversion units which the first NGNSP's have called CSLAMs* or MSANs.† TISPAN (see Section 3.5) defined them as Access GateWays (AGW), located at the NGNSPs PSTN Emulation Subsystem (PES). These units provide analog lines, ISDN basic/primary rate interfaces and/or V5 interfaces (toward remote subscriber units) to residential end users and enterprises; and they speak the NGN language (SIP and RTP) to the NGN core. They are today outside of the scope of IETF and 3GPP standardization, but widely available. Again, the question is whether the new NGN provider would operate those himself or peer with specialized TDM access providers via IP (SIP and RTP).

Peered with IP Access Networks such as DSL, cable, or WiMAX, through which SIP signaling and RTP media flow directly to the User Agents, over IP; the combination of such next generation Access Network and Core Network is sometimes referred to as the "All IP network." The layer-2 connectivity to the end user (e.g., PPPoE: Point-to-Point Protocol over Ethernet, carrying IP) is offered by Broadband Remote Access Servers (BRAS) — the last element of the access network.

Peered with Enterprise IP Networks, either already operating their own IP PBX, or using media gateways in front of their circuit-switched PBX. These media gateways were originally deployed by the enterprises,‡ but there's a trend to outsource this to the NGNSP§. In another scenario, the NGNSP operates a shared "IP Centrex" service platform, to which different companies' user agents (fixed and wireless IP phones, user agents on laptops, WiFi PDAs, etc.), are registering via SIP or the more feature-rich (but Cisco-specific) SKINNY Client Control Protocol (SCCP). We mentioned this as an interesting opportunity back in Section 2.3. It is then up to the IP Centrex service provider to keep "Berlin Walls" between the user data of different enterprises. However, many communications between users of different enterprises could be routed locally within the IP Centrex platform — that's a benefit.

Peered with other NGNSPs, including long-distance VoIP carriers. The name of the game here is currently "least cost routing," but this could rapidly evolve to "known quality routing." The new Service provider.

It is very likely that specialized TDM trunking providers and TDM access providers, or at least dedicated Business Units, will appear over the next 5–10 years. Investing in circuit-to-packet gateways only, without the costs and risks of an NGN

---

* Combined Subscriber Line Access Multiplexers.
† Multi-Service Access Nodes.
‡ Thus to be considered as a TISPAN Residential GateWay (RGW).
§ Thus to be considered as a TISPAN AGW.

core network and its set of VAS platforms, is actually a very attractive business model for today's IXCs* and ILECs.

If that happens, new NGNSP's will be able to use the same *NGN peering* equipment, processes and contracts with TDM trunking providers, TDM access providers, IP access networks, enterprise IP networks and other NGNSPs.

The key equipment you need for NGN peering is a Session Border Controller (SBC). According to the TISPAN definitions we reviewed in Section 3.5, an SBC performs both the I-BCF and the I-BGF, at the edge between the core network and other NGNSP's. At the edge between access and core networks, TISPAN defines the SBC as performing the P-CSCF and CBGF.

But, following the idea of generic NGN peering equipment, there's no reason to distinguish these two SBC deployment cases anymore.

In an open world, there's no reason to consider the access network "trusted" and peer NGNSPs as "untrusted."

The User Agents accessing the NGN Core Network (end users in an IP Access Network or an Enterprise IP Network; an SBC or SIP proxy server in a peer NGN) may never be "trusted":

> They might be tempted to set up a session and disconnect it immediately, only to discover the RTP ports, and continue to exchange media on these ports for a longer period, after the end of the initial short session.
>
> They might be tempted to pass payload (text, images or even audio and video) within the signaling messages; SIP is a very extensible protocol and this is perhaps also its weakness.
>
> They might try and discover the network topology, to launch malicious attacks such as SIP Denial-of-Service (attempting to set up a huge number of SIP sessions without ever actually completing any).

4.7   Session Border Control

The Session Border Controller (SBC) is the element that allows an NGNSP to securely interconnect their core network to access networks, IP-enabled enterprises and peer NGNSP.

It is thus the instrument by excellence to realize NGN peering — well, at least the Session Control and the Border Gateways layers.

Ideally, an SBC should be decomposed into a Border Control Function (BCF) in the session control layer, and a border gateway (BGW) in the media and border gateways layer, interconnected via the ITU-T H.248.1 gateway control protocol. We looked at that architecture back in Section 3.1, and at the protocol in Section 3.6.6.

---

* International eXchange Carriers, operating circuit-switched international exchanges, and well-interconnected to ILECs and CLECs.

The SBC plays important roles: media cut-through, NAT traversal for signaling, media stream latching, topology hiding, transcoding, admission control and more than a dozen other functions.

As most of the functions rely on tight integration between the BCF and the BGW, there's today a technico-commercial debate on whether the BCF should reside inside the NGN core network softswitch (a server), or may be colocated with the BGW (a router).

4.7.1   Media Cut-Through

As soon as third-party IP-based networks need to be connected to an NGN or IMS, thus in the latter three cases of the previous section, an SBC is required to prevent media transfer outside of correctly established, chargeable SIP sessions.

More precisely, during session set-up, early media (ring-tones etc) may already be received by the session originator. The ring tones may be generated by the responding user agent or by the network (e.g., by a media server). But the forward audio (and video) channels should be opened only when a SIP 200 OK message is received from the responding user agent.

This process is called "media cut-through" and is the SBC's primary task.

4.7.2   NAT Traversal for Signaling

A second role of the SBC is to help SIP sessions to be established through enterprise FireWalls (FW) and Network Address Translators (NAT).

Different types of NAT exist, but the most commonly used form is symmetric NAT, in which all requests from the same internal (private) IP address and port to a specific destination IP address and port are mapped to a unique external (public or private*) source IP address and port. If the same internal host sends a packet with the same source address and port to a different destination, a different mapping is used.

With Simple Traversal of UDP through NAT (STUN), a user agent inside the enterprise environment (containing a STUN client) can ask a STUN server on the external network to tell him what the external address of the NAT is, and what port was opened by the NAT to allow incoming traffic back in to the network. Obviously, the STUN server needs to be colocated with the NGN operator's SBC. Otherwise, the symmetric NAT will allocate different external ports for the STUN query as for the upcoming SIP request.

The originating user agent can then use that STUN response to fill in the IP address in the From: field of the outbound SIP INVITE (or other message). This will allow the first SIP proxy (the service provider's SBC in this case) to return SIP responses to the originating user agent, via the NAT's external address and port.

---

* The NGN provider may, on the interconnection to a corporate network, have obtained an address from the private address space of that corporate network. An entire private address space may even be used on the interconnection only, with NAT functions both on the service provider and the enterprise side.

For secure communication to the NGN's SIP proxy, the originating user agent may eventually encrypt that outbound SIP INVITE message using IP secure (IPSec).

If no IPSec is used, the enterprise Firewall's Application-Level Gateway (ALG) for SIP is able to look at the requested RTP ports for media, and may open them on-the-fly — whereas normally they are kept securely closed.

If IPSec is used, the enterprise ALG is unable to decrypt the SIP message, so it cannot open RTP ports on-the-fly. The originating user agents must thus use RTP ports which are preconfigured as open on the Firewall. If the enterprise's IT department is not willing to leave dedicated ports open, IPsec will have to be sacrificed.

While STUN allows an originating user agent on an enterprise LAN to set up outbound SIP sessions through a NAT, clearly it doesn't allow the external world (the NGNSP in this case) to set up an inbound session to that SIP user agent. The port (on the external side of the NAT) on which SIP INVITEs should be received keeps changing, whenever the user agent obtains a different internal IP address, via Dynamic Host Control Protocol (DHCP). In most enterprise networks, new IP addresses are allocated every couple of days.

Traversal Using Relay NAT (TURN) is a principle that allows the SIP User Agent to obtain a *permanent* external address and port from a TURN server on the external network, at which the UA can then be contacted.

From the description above, it would sound as if SIP user agents should absolutely perform a STUN or TURN query before registering to an NGN and announcing their Contact: address, or would otherwise be unreachable when in the enterprise environment. STUN and TURN servers should reside inside the SBC that is going to be used to deliver that inbound communication. Otherwise the information is unreliable, certainly if the enterprise has multiple NAT boxes and ISPs.

As most SIP user agents are today still shipping without STUN/TURN client, this issue could hamper the deployment of the NGN itself, until each of these devices can be allocated a permanent (static) IPv6 address.

The situation is not that dramatic.

Today's smart SBC's work in Application-Level Gateway (ALG) mode, for user agents unable to perform the STUN/TURN queries. These smart SBC's automatically translate the periodic SIP registration requests to include the temporary external address and port on the NAT box.

If the SBC notices that the user agent's IP address is frequently changing (in SIP registrations and session initiation requests), thus that the external address and port allocated by the NAT are unreliable, the SBC will even insert one of its own permanent IP addresses and ports in the SIP registration, and act as the TURN server.

4.7.3 Media Stream Latching

All the above helps the SIP signaling to flow in the two directions, between user agents behind NAT and an NGN. How about the media streams themselves?

When an originating user agent sets up a session (SIP INVITE), it includes its own address as the Origin ("o=" parameter) of the Session Description Protocol (SDP) offer. If it hasn't performed any STUN/TURN query, that address is internal, unreachable by the NGNSP, for downstream media toward that user agent. If the user agent did the STN/TURN query, the address in there might be good. In general, from the NGNSP's perspective, it is thus unreliable.

Media latching is a principle whereby the SBC ignores that address information in SDP, but monitors the source IP address and port of the upstream RTP packets, in order to determine the destination IP address and port of downstream RTP packets. It is a continuous process, not only occurring just after session establishment.

To speak in TISPAN and H.248 terms, the Border Control Function (BCF) should be able to instruct the Border Gateway Function (BGF) to ignore the address it sends in the H.248.1 ADD request, but to rely on media stream latching instead.

4.7.4   Topology Hiding

An SBC should also include a Topology Hiding Internetwork Gateway (THIG).

This function makes sure that the list of hosts through which the SIP request is proxied (in the Via: and Record-route: headers) is erased from the SIP header before sending out the message to the outside world (untrusted or trusted access network, enterprise or peer network).

This protects the NGN from Denial-of-Service (DoS) and Distributed Denial-of-Service (DDoS) attacks that could be set up against the NGN core network elements and VAS platforms.

4.7.5   Transcoding

As mentioned in Section 3.6, the ITU-T and ETSI/3GPP have defined several audio and voice encoding formats.

We looked at some of them back in Section 3.6.4.

The SBC is ideally placed to transcode the formats into each other. For example, optimal audio quality might be realizable with the G.711 codec (64 Kbit/s) on the NGN provider's own backbone, but a good quality/price ratio might be obtained from long-distance providers using G.723.1 at 5.3 or at 6.3 Kbit/s.

Ideally, the SBC would also perform transcoding for video formats and media types other than voice. For example, today 3G circuit-switched video calls use the H.324M codec, whereas most "fixed" SIP-based video sessions use the H.263 or H.264 codec. If mobile users wished to participate in enterprise video sessions one day, transcoding would be required.

Transrating is the process to adapt the packetization interval — for example, from 20 ms of encoded voice in the RTP payload, to two RTP packets of 10 ms each. Indeed, the standards organizations, national regulators, service providers and enterprises do not always seem to have come to compatible choices in that area either.

4.7.6 Admission Control

SBC's are also the ideal place to limit the total number of sessions being admitted from a given source domain, in order to protect the NGN core network from overflow.

In a network with multiple SBC's across the globe, it becomes difficult for these SBC's to share a common set of real-time counters, representing the total number of ongoing sessions to/from a given NGN peer, enterprise or application service provider.

Hence as we saw in Section 3.5, the more recent view is that the SBC, more specifically its interconnection border control function (I-BCF), should query a central Admission and Resource Control layer (RACS).

Indeed, the RACS has an overview of the total number of sessions going on in the whole network, and can take into account the various interconnection agreements.

The need for admission control is also related to the need for overload protection; we'll look at that in Section 6.11.

4.7.7 Other Functions

The SBC is such an important and strategically positioned new element in a NGN architecture that new functions keep being added to it:

Lawful Interception (LI), where certain media streams need to be intercepted and copied to the law enforcement agencies

Virtual Local Area Network (VLAN) tagging, allowing it, for example, to carry signaling and payload on separate VLANs

Quality-of-Service (QoS) marking

QoS and flow statistics reporting

Bandwidth policing, probably by querying the TISPAN RACS we saw in Section 3.5

Echo Canceling

DTMF interworking between multiple methods, in-band* and out-of-band[†]

SIP security, screening the use of SIP messages, header fields and values, for each interconnected peer NGN or enterprise

Accounting

Least-cost routing

Service assurance

Reporting on media quality

Some of the functions (e.g., QoS marking) rely on tightening the bonds between the media and border gateways layer and the IP transport layer.

---

* Using RTP payload or RFC 2833.
[†] Using either SIP INFO or NOTIFY messages.

Leading suppliers and more than a dozen start-up companies have succeeded to build and commercialize these platforms, which are now commonly considered as the "edge routers" and "firewalls" of Next Generation Networks.

4.8 ASP Interconnections

In the past, various standards had been proposed to open a service provider's network to third party Application Service Providers (ASP).

More specifically, the standards bodies launched open Application Programming Interfaces (API's) that could be offered to ASP's, allowing them to use the Service Capabilities ("the bottom part") of the VAS platforms.

The common point of these standards (e.g., Java J2EE, Parlay, Parlay X, OMA) is that they are very focused on the mechanics of call control, but overlooked essential aspects such as subscription (which ASP may interact with which user), inter-operator accounting (how would revenue be shared) or media (who would play announcements, ring tones, videos, etc.).

Therefore, they appeared to be very unsuccessful at opening up PSTN and GSM networks to developer communities.

In the future, VAS platforms may of course be further developed to offer access to their service capabilities, by third party ASP's, for charging or content delivery purposes. We'll take a look at this strategy in Sections 5.12 and 6.18, respectively.

On the other hand, GSM service providers have been successful at building up a network of premium SMS service providers (e.g., SMS "Large Accounts," "Short Codes") via pretty basic TCP/IP protocols*. These SMS service providers are semi-trustable; constant monitoring is required to protect the GSM end users from fraud, scams and spam.

Can the NGN be opened up in a similarly simple way, but including better security, in order to connect more ASPs in the long term?

When SMPP was conceived, the GSM network was not accessible by IP.

But fundamentally, in an NGN, where IP access is ubiquitous, is there still a reason to make a distinction between an end user and an ASP?

End users are, for example, admitted to a 3GPP IMS based after ISIM authentication and authorization based on a profile held in the HSS (see Section 3.3.2).

And as we saw in the previous section, enterprises, access networks, and peer NGNSP will be admitted to the NGN core network by SBCs.

If we want to avoid developing yet another access method, besides end user access (via IMS P-CSCF) and SBCs (TISPAN I-BCF), should ASPs then be treated as end users, or as peer networks?

---

* For example, Short Message Peer to Peer, developed by a little Irish company Aldiscon, later acquired by Logica, now LogicaCMG; in 1999, SMPP was handed over to the SMS forum.

Of course, ASPs are in a special position. Contrary to peer NGNSP's, ASP's are proposing their services to the NGNSP's subscribers (e.g., individuals, enterprises). ASPs will have specific requirements such as the possibility to do the following:

Offer free services to the end user, services which are then perhaps funded by advertisement or other business models (see also Section 5.6) or otherwise announce the price of an item to the end user (Section 5.7)

Obtain the end user's agreement

Check an end user's account state, and perhaps reserve funds, prior to delivering the service or item

Assess whether the service or item was properly delivered

Debit the user's account when this is the case, and notify the end user

Request the NGNSP to waive other charges associated with the delivery (i.e., the charge for the transported volume in Megabits, associated to a game level)

Include ASP services in the NGNSP's product catalog, so the end user bill (or online log) would contain an accurate description of the services rendered and items purchased

Participate in the NGNSP's subscription, lawful intercept and customer care processes

The requirements go far beyond the mechanics of session control and media injection.

Therefore, ASPs should be treated as interconnection partners rather than end users.

End users should continue to be authenticated and granted access via standard mechanisms (e.g., SIP, IMS), not "polluted" by ASP requirements.

The NGNSP should offer an open interface on their BSS. That new, online, transaction-based BSS not only needs to charge for end user communications, but should become a BSS for Wholesale, ASP and Enterprise billing. These third parties should obtain a secure API to perform the account management and charging interactions listed above (e.g., check an account state, reserve funds).

As for the delivery of the content, the ASP should set up a session via the I-BCF, the top part of the SBC which will already provide security, bandwidth policing and other functions. That I-BCF may now need to link to the BSS in real time, to inform the BSS of successful service delivery.

In summary and according to Figure 4.4, ASPs could be granted access to the following:

The Service Capabilities sublayer of the Application Servers, which would offer the same capabilities to third party applications as to the NGNSP's own applications; this method requires further development well beyond the current standards; we'll also mention this need for openness in Section 6.18, when we'll list the requirements for VAS platforms in general

The NGNSP's Billing Support System (BSS), via an online API offering the ASP access to account management and charging operations, also in a secure, subscription-aware manner; we'll look at this in a wider context in Section 5.12

The Session Control layer (namely, the TISPAN I-BCF from Section 3.5), where ASP's can accept and set up sessions to the NGNSP's end customers, subject to access control and resource allocation by the NGNSP's Resource and Admission Control plane

The Media and Border Gateways layer (more specifically, the TISPAN I-BGF), where they can inject media such as audio, video, Web sessions, in a controlled and metered way

And if these capabilities are developed for the ASPs, nothing would prevent an enterprise to benefit from them, for example, to obtain a real-time view on the upcoming next invoice, to create new end users, organize them into separate accounts, or to order new services — or, to take an example involving session

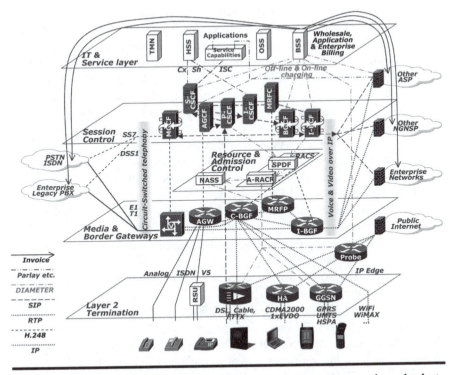

Figure 4.4 Wholesale and retail charging for NGN services and methods to open the NGN to third-party Application Service Providers (ASPs).

control, resource control, and border gateways, to set up an executive video communication to the employees.

Last but not least, the diagram also highlights the fact that, besides offering new online transactional capabilities, a BSS needs to continue invoicing end users but also third parties (PSTN/ISDN, Enterprises, Other NGNSP and the ASP) for the usage of the communication infrastructure.

4.9 Conclusion

We are arriving at the end of our chapter, so let's summarize our reasoning and findings, while trying to engineer our NGN.

We started from a long list of questions to be answered in the business plan of ngvas.com, our hypothetical little NGNSP.

We discussed the need for NGNSPs to advertise their media quality and price toward certain destinations, hence to be automatically discovered by other NGNSPs and end users, in a worldwide jungle of competitors.

This situation calls for an advertising overlay, an independent organization in which the quality metrics and price information could be exchanged in real time — the "stock exchange" function of the NGN.

With business models moving from retail to wholesale, NGNSPs will have to use a new form of subtle, thin distribution, which boils down to convincing the established IM players to add the standard protocols in their client software. Indeed, blunt linking of an IM client or device to a given service provider — linking iPods to iTunes — will not be tolerated by the market much longer.

An idea for differentiation is for NGNSPs to focus either on outbound communications (thus charge end users and enterprises), or on inbound communications (thus live from interconnect revenue only).

The cornerstone of each NGN strategy should be NGN peering, rather than TDM interconnection. TDM trunking and access are best left to specialized players — probably today's ILECs and IXCs.

A new, dedicated network element has appeared for this purpose — the Session Border Controller.

SBCs could also play a role in allowing third party Application Service Providers to be interconnected to the NGN, though other methods exist.

Considering the blurred distinction between ASP and other peer NGNSP, and the need for a uniform wholesale charging policy for both third party NGNSPs and ASPs, the SBCs are also the best option to interconnect ASPs.

Finally, the ASP requirements we examined in this chapter are a first indication that we should take a closer look at the charging and rating requirements for new communication means, which is the topic of our next Chapter 5.

After that, we'll take a dive into the technical architecture of VAS platforms, to engineer the Service layer properly in Chapter 6.

Chapter 5

Charging and Rating Requirements for New Communication Means

As the CFO of our ngvas.com NGNSP would say, of course it is required to develop tariff plans and payment methods, for either true communication services or for the transmission of stored media.

But tariff plans alone should not become the primary concern of a communication service provider's marketing department. The excess of creativity in this area results in a general sentiment of unfairness and alienation, rather than in a trusted and valued brand. The initial early adopters do not switch to the new tariff plans, promotions and payment methods, and feel left behind.

And the name of the game is customer ownership. Once this high quality customer relationship is created (the "account"), several services may be sold through it: connectivity, communication services, and even content, physical goods and other services than communication.

Inventing and implementing charging and billing schemes for too many of these could eventually again turn into a digression.

The purpose of this chapter is not to make a sterile overview of charging and rating systems used by today's communication service providers, nor to try and compare them or provide migration advice. The fact that the world is moving to real-time rather than batch processing, for example, is an example.

Let's first try and build a set of essential definitions, and then examine the likelihood for future communication means to require a product catalog, charging, rating, metering, advice of charge, redirection, billing, promotions, openness and/or correlation.

5.1 Product Catalog

Since the mid 1990s, mobile service providers have multiplied the number of new "product" offers — merely corresponding to handset subsidies and tariff plans, rather than to packages of value-added services. It is as if petrol companies would have subsidized new cars, and invented various packages containing a monthly mileage or a number of gallons of fuel. Plus a higher price per extra gallon, plus a more expensive rate if you would fill your tank at competing petrol stations (roaming).

New, more competitive product offers were launched to grab extra market share, without inviting the earlier subscribers to migrate from the old products. This created a situation with sometimes more than 100 tariff plans on a single network.

In such an environment, it is impossible to launch any new service to the existing subscriber base. Any new service offer results in doubling the number of products, as each product now exists in a version without and with the new service.

A well-structured product catalog should be at the heart of any modern communications company. All processes within a communication service provider (rating, charging, billing but also marketing, retail and customer care) depend on it, and should interact with it in real time. The accuracy and quality of these processes will depend on the quality of the product catalog.

From experience, communication service providers have converged to the following elementary object model of a product catalog. Single-headed arrows represent one-to-many integrity relationships, double-headed arrows are many-to-many relationships:

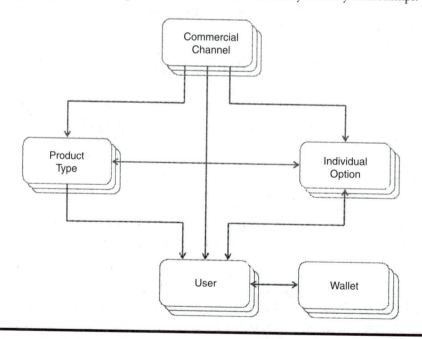

Figure 5.1   A simple object model of a product catalog.

A product catalog defines the offer to the end user, in terms of product types (basic and advanced services) and additional individual options.

The product catalog needs to include information on the compatibility of these individual options (mutual and with the product types). This is illustrated by the many-to-many relationship between product types and individual options. One product type may include multiple individual options, but a particular option may occur in several product types.

The product catalog (product types and options) could be different for each commercial channel — each organization (e.g., residential/business division, reseller, internal test lab) selling communication services to the end users, even if technically all channels are hosted in a single environment. Such system is also called "multitenant."

However, having a dedicated product type for a single customer could be one bridge too far, even toward a large corporate organization. It is then probably better to grant tailored discounts, or maybe promotional rates for on-net communications inside a virtual private network.

Typically, the tariff being applied (e.g., activation fees, periodic subscription fees, source-destination combination*, measurement unit†, initial charge, subsequent charge per unit) will depend on the combination of all objects in the product catalog:

> By which commercial channel was this user acquired?
> Which product type did the user select (e.g., gold, silver, vip)?
> What individual options did the user subscribe to (e.g., a promotional messaging bundle)?
> Which wallet is the user currently using (e.g., an company's shared wallet in the daytime, or an individual wallet at night — see Section 5.2)?

It is thus wrong to attach a tariff plan only to a Product Type.

Over time, a product catalog is a living database and should therefore be fully versioned, indicating the date and time of start and end (past and future dates) of each aspect of the offer, however tiny. When any offer ends (product type, option), it should specify the new offer by which it is replaced (and to which users are thus going to be migrated). This is often forgotten in the design of basic charging systems.

How complex should a product catalog become? In marketing terms, we are looking for the perfect match between a limited number of appropriately positioned products (+ price, place and promotion) for each customer segment (defined based on their usage of the communication means). Besides the limited number of products (e.g., Silver, Gold and VIP), additional individual options may be purchased or simply selected by the customer (e.g., an insurance for the cell phone, a favorite destination country, or friends and family numbers with lower rates).

---

* From my home zone to a premium rate video site.
† Seconds, minutes, megabits, Web pages, game levels, and so on.

Within communication service providers, but in general within many commercial organizations, there's a difficult equilibrium to be found between the mix of product types and the tailoring options. Having too many product types and no individual options usually has a greater impact on the organization (in terms of complexity, cost, training, etc) than having too few product types and too many options.

Ideally, the product catalog should regularly be compared to the subscriber database in order to detect and suppress unpopular product types and options, hence to simplify the offer toward the end user. We'll talk about this in the context of reporting (Section 6.25).

In the peer-to-peer world, the product catalog should be part of the user interface or at least of some central Web site explaining the offer to the peer-to-peer users.

Subscribers and prepaid users of telecom services used to have a lifecycle: preuse, active, dormant, suspended, fraudulent, porting out pending, and so on — please refer to the Section 6.2 on user accounts. Any lifecycle should not also include the current state but the history and future planned states, as of precise dates and time. A new evolution is that product catalogs are evolving to include a lifecycle and renewal policy also for the product type and for the individual options.

For example, the user subscribing to the cell phone insurance could select to have this personal option starting on the first day of the next month. The option could be automatically renewed each month if there are sufficient funds in the account, or the system could automatically ask for confirmation prior to renewing that option, or the user could opt for a scheme in which the option is suppressed after one month — for example to buy the insurance only to go on holidays.

The tariff implications of that renewal policy is an area that is going to be developed in the coming years.

5.2   Account = User + Wallet

An account is a term for a customer relationship with an end user of communication means, but may also be used w.r.t. a revenue sharing partner (e.g., carrier, roaming partner, application service provider).

Some communication users could demand multiple accounts, even for the same communication events. For example, corporate organizations today want to look at their end users and usage by division, by country, by communications product and worldwide.

The degree of trust should determine the prepaid or postpaid nature of such account, more than type of user or partner. For example, when NGNSP will have thousands of NGN peering relationships, we expect prepaid accounts to appear also for (what is today called) inter-operator accounting.

What is stored in such account is credit, expressed either in monetary or in nonmonetary *balances* (e.g., minutes, megabits). The balances in an account may be of the debit type (positive values, "prepaid") or credit type (negative values, "postpaid").

Limited credit and single-use (non-rechargeable) balances are the two other types. An account will thus never be 100% prepaid or 100% postpaid.

A single balance may be organized in multiple *buckets* of credit, each with a different validity date (after which the credit is suppressed) or future installment date (as of which the credit becomes usable). A balance thus has no expiration date; it is the bucket which has it.

A good idea is also to group multiple balances into a wallet. An end user could use credit from multiple *wallets*: a shared company wallet for business communications in the day time, a shared family wallet for international calls to a favorite destination, and a personal wallet for all other calls, for example. The wallet to be used could be determined either by the end user or by the communication service provider, based on the service being used, the destination, time of the week, and so on: a real "wallet usage policy."

What kind of data will we find in the *user*'s profile? In there, for end users as well as partners, we will find the user's identification and contact details, the history or customer care contacts, the lifecycle data, links to purchased options (with their own lifecycles and renewal policies), preferred destinations (e.g., friends and family) and negotiated discounts.

As the user and the wallet together determine the tariff being applied, besides of course the commercial channel and individual options which nevertheless have less effect, the combination of a user and a wallet forms a valid definition of an *account*.

An account is thus not only a user having selected a product type and options, nor only a wallet full of balances and buckets (credit). It is encompassing both aspects.

An account is also not to be confused with a *payment channel*. Multiple payment channels (an invoicing address, a credit card authorization, a scratch voucher, a bank-to-mobile transfer, etc.) could indeed be used to settle a bill or top-up a wallet. A scratch voucher could for example be used to settle a postpaid bill.

Mobile operators used to implement the user account databases as "Operator Support Systems" (OSS) on one side and "Prepaid Intelligent Network" systems on the other, but there's no reason to make such distinction. A generic "accounts system" would do, and could even be used for reseller/dealer accounts, inter-operator accounting and application service providers.

Are the options specific to a user? Today they are, but one could imagine options on the wallets too. In that sense it is a more future-proof idea to link new options to an account rather than to a user or partner profile.

5.3 Charging

Charging is the process through which the network delivering the communications service is interacting with the system holding the accounts.

In theory, the network should interact with a charging system, which in turn should tap from an accounts system, which would finally consult a product catalog.

In practice it is not easy to define open interfaces between these, so the three functions should remain combined in a same "charging system."

The timeliness and speed of this interaction is described in terms as "real-time," "hot," "warm," "cold," but the truly important distinction is "online" versus "off-line."

In online charging, the network consults the charging system prior to delivering the service, and asks for permission and for credit (in nonmonetary units).

In off-line charging, the network informs the charging system that the service just got delivered.

Historically the online approach has been more suitable for user charging (retail), as it avoided any fraud in conjunction with prepaid payment methods. With off-line charging, prepaid services could be delivered while the charging system would notice afterwards that the balance was too low to pay for these. The services (calling, texting) would then have to be deprovisioned in the network, and reprovisioned after top-up.

And the off-line approach was preferred for inter-operator accounting (wholesale), as the network could pass more information to the charging system (trunk group being used for the outbound or inbound call) than through the online charging interfaces.

This has lead to a few industrial camps and matching departments within the operators: "Service Nodes" / "Prepaid IN" for online charging; and "Billing Mediation" + "Billing Support Systems" for off-line charging.

Today there's no technical nor economic justification anymore to use off-line charging for postpaid (contract) users — except maybe risk sharing. Online charging systems can and should charge every user whether prepaid, postpaid, credit limited postpaid or single use (e.g., a calling card).

The online charging approach remains the ideal target method if it could address inter-operator accounting.

However, today we see little evidence that the online charging interface of the next generation network will support this.

5.4   Rating

Rating is the process of letting the charging process interact with the accounts system and the product catalog to determine the price of a communications event.

In general, it happens within the communication service provider's online or off-line charging system. Alternatively, a third party could perform rating if it has its own product catalog (see Section 5.12) but let's for the moment set aside this possibility. All interaction between the network and the charging process is based on nonmonetary units.

As an online/off-line charging process, a product catalog or an accounts database cannot be described in a generic way, it is also difficult to describe the rating process.

For online charging, at some point the charging process reserves a slice of credit out of a balance (or multiple balances in a cascade) — a process known as credit slicing. Ideally, the slice of credit is small enough to permit multiple simultaneous services to be delivered with that balance, but large enough not to lead to too frequent charging interaction with the network. The charging process needs to consult the account's product type in order to know which balance cascade to tap from.

If the balance is monetary, inverse rating takes place: the inverse rating process consults the product catalog and determines the total price or the price per unit of the communication event. It transforms the credit slice from monetary to nonmonetary units.

Let's stress that, if all balances were expressed in minutes, no inverse rating would occur in order to charge online for calls!

If credit reservation took place, the online charging process then allocates the number of nonmonetary units (e.g., seconds, megabits, clicks) to the network or just gives an acknowledgement ("go" signal). In some implementations of time-based charging, the online charging system just gives a go signal but does not instruct the network to meter the seconds. It runs its own timer and will either tap more credit out of the wallet, or interrupt the call if there's no credit anymore.

The off-line charging process usually does not acknowledge the request for debiting. Rating may occur immediately or much later: looking up the service in the product catalog and transforming nonmonetary into monetary units.

For off-line charging as for online charging, the charging process ends up debiting (lowering) the balance (monetary or not).

In conclusion, off-line charging uses rating, and online charging uses inverse rating, only if the balance is monetary.

If, in a NGN, online charging thus became possible for inter-operator accounting, and all balances were expressed in nonmonetary units, no rating would be required anymore!

5.5 Metering

As described above, the online charging process may allocate the number of nonmonetary units to the network.

Usually these used to be the seconds granted by a prepaid IN system (i.e., an online charging process) to a Mobile Switching Center (MSC).

In the last 5 years, however, advanced metering devices have been developed that are able to meter the seconds, kilobits or number of clicks for a wide palette of services and destinations, even within a single mobile data session.

Upon establishment of such session, the metering device retrieves the palette of services to be metered, for a given user, from a central Authorization server. The metering device is also informed whether online or off-line charging is to be applied for each service and/or destination. Some services/destinations could be marked as forbidden (black-listed) or sabotaged (gray-listed; e.g., competitors' applications). Also, it is made aware what to do with traffic that cannot be recognized as being part of any service/destination ("background traffic").

In the coming years, the palette of services to be metered should probably be merged into a single product catalog. Too often, the definition of available data services have remained separate from the main voice and messaging product catalog.

If the background traffic requires online charging, the metering device obtains a credit slice from the online charging system.

When the user/partner accesses a service/destination defined in the palette, the metering device obtains credit (sometimes also referred to as "quota" in this context) from the online charging system, or produces tickets for off-line charging.

Highly granular metering is thus required for both interaction with online charging systems, as well as for off-line charging.

It is sometimes referred to as "content charging" but that's an inaccurate description, as content can be charged without metering, by applications (end points) tapping directly from the account (see Section 5.12 on openness).

5.6   Reverse Charging

Most users understand the principle of a toll-free (800) number: the call is free to the caller because it's fully paid by the called party.

The decision for the charging system to allocate the charges to the content service provider (the freephone number owner in this case) used to be simple (namely based on the called number range).

That is changing in modern communication services.

Let's consider the example of an online game where the game itself is charged per new downloaded game level, and normal GPRS volume charges apply for the nondownload traffic.

The metering device is smart enough (or instructed) not to take into account the game level downloading packets to calculate the volume of the background session — and to modify the background session to become time-based when using WiFi.

If a new game level download succeeds, it is charged to the end user as a unit price.

If the user had already successfully downloaded that game but later lost or erased it, the game level can be restored and the gaming user pays only for volume charges — except when on WiFi.

If, however, a game level download is incomplete or unsuccessful, the volume corresponding to that failed download should be charged to the game provider. Again, except when on WiFi.

This small example illustrates that charging decisions are far from being static, and that the communication service provider's charging processes should be in full control of normal vs. reverse charging.

5.7 Advice of Charge

As the complexity of the product catalog and account structure increases, so does the need for clear Advice of Charge (AoC) to the end user, and sometimes explicit acknowledgement or even additional authentication, in advance of actually delivering the service.

In PSTN, toll exchanges completing premium rate calls are able to inform the originating local exchange of the tariff in real time (through signaling). The originating local exchange can then inform payphones of the charges to be applied (16 KHz pulses).

In some countries it is a legal requirement to inform the customer of the tariff of premium rate calls, premium SMS, or to inform the caller that the destination number is ported out (with a warning beep).

Metering devices are also able to redirect the mobile data request to an AoC page, or to insert the AoC information in the downlink content.

What will be the price of SIP sessions to various destination domains, worldwide?

Increasingly for data services and next generation services, the operator will be held accountable for informing the users and partners of costs, in advance of delivering the actual service.

In the NGN standards there's currently no sign of an AoC facility.

Early adopters won't bother, but cost conscious users might hesitate to use the services.

5.8 Redirection

As mentioned above, mobile data users are sometimes redirected to AoC pages or popup windows and may explicitly need to agree.

In case the metering device does not obtain credit for a service to be charged online, or in case the service is barred, it should be able to redirect the user to an explanation page (autonomously or under instruction of the online charging process).

If the problem is an empty purse, the user should be invited to top-up the credit, and in case of bad debt, to pay the bills.

Maybe in some extreme cases, redirection to a human CRM agent (via text, voice or video call) would help?

5.9 Billing

Billing is the process of producing a bill, an invoice.

The input for billing is a set of rated CDRs (i.e., already containing a price) coming out of the charging processes.

Today, most operators generate invoices for only postpaid accounts (charged in off-line mode). Prepaid subscribers can get an invoice for top-up events (e.g., for vouchers) but not for usage (calls, messages).

But there's no technical reason to link the production of a bill to the charging mode — that is, billing should be possible in case of online charging too. Prepaid services could appear on an invoice, that is, to allow the customer to pay VAT only on usage, rather than today on the facial value of the top-up vouchers. Top-up vouchers could then be sold without VAT, to customers having a verified billing relationship (name and address).

As for product types and lifecycles, one can wonder whether bills should not be produced regarding a wallet, rather than regarding a user or partner.

5.10   Payment

The payment methods should not be imposed by the bill. It should be perfectly possible to settle a postpaid bill using a prepaid scratch voucher, ATM reload or bank-to-account or inter-account credit transfer — whereas these payment methods are today considered only to top up prepaid accounts.

As for some public utility services, the customer could be proposed the option to pay a monthly fixed amount, to even out consumption peaks. The monthly amount would be recalculated annually.

As for tax payments in some countries, the customer could be invited to make advance payments in return for lower rates or bonuses (see Section 5.11 on promotions).

Credit notes should be issued to correct billing mistakes — nobody's perfect.

Customers and partners having real payment problems should be converted to prepaid or at least limited credit accounts.

Bills, credit notes and payments should be tracked in a ledger closely linked to the account, for customer care reasons.

Payment chasing represents today roughly one third of an average communication service provider's CRM department.

5.11   Promotions

Promotions can be granted upon subscription, upon usage and upon payment.

In the early stages of the development of a communication service, when it is important for service providers to grab new users as fast as possible and churn doesn't matter, promotions are mainly granted upon subscription. In more mature markets, service usage and payments (e.g., top-up of prepaid account) are taken into account to stimulate loyalty and reduce churn among the most profitable users.

Promotion systems calculate the eligibility of end users and partners, based on multiple factors (e.g., address, lifecycle, usage, top-up behavior), but knowing the customer segments certainly helps. It is of no use to grant free MMSs to the Silver Cynics; they'd, however, very much appreciate free texts to their grandchildren.

The system ends up applying a promotional action: moving the account up the ladder in a loyalty scheme (e.g., from Gold to VIP in a frequent caller program), granting bonus credit in the wallets, and so on. Please note that it is often the wallet receiving this promotion, not the user/partner itself.

A promotion loses its effect if the users of the wallet are not informed in near real-time of it. Granting a bonus at the end of the month will have only a minor effect on loyalty.

In most cases, but certainly if a promotion is granted upon subscription to a service, loyalty can be greatly improved by not granting the whole reward at once. Credit bonuses can be spread over time as "periodic credit installments" (e.g., one free SMS a day for the coming month).

It is important to be able to simulate a promotional campaign in advance of actually launching it. Too often, communication service providers launch campaigns with meager eligibility criteria, to unclear segments (too many people) and granting an insignificant or inappropriate benefit.

Promotions can be used as a tool to steer end users toward more profitable behavior, for example to use e-vouchers rather than scratch vouchers to top-up their prepaid wallets, or to top-up with larger amounts rather than frequent small amounts.

5.12 Openness

Much has been said and written regarding the edge between mobile operators and financial institutions.

Operations on a mobile account (e.g., topping up a prepaid account using a scratch voucher) are for example subject to VAT, contrary to a bank account. A communications invoice is sent to you including VAT.

M-commerce has been touted as the possibility for the mobile communications consumers, using their mobile account, to purchase various services and goods, other than mobile communications and content, sometimes forgetting about these VAT issues.

Gradually, the mobile service providers realized that by opening up that "purse" for a competitive fee (meaning at the level with credit card fees), they would face competing services being offered by these third parties directly to their own subscriber base.

We looked at this from a technical angle in Section 4.8.

Premium SMS and MMS (short messages and multimedia messages to short codes) have been offered as means for third party application service providers to "tap" (debit funds) from the user's prepaid mobile account, or to add items on the postpaid bill. It is a case of bundled selling, where the mobile service providers do not wish to offer access to the purse (the payment method) without a message being carried over their networks (the communication means and connectivity).

If the payment method was offered stand-alone, then premium SMS (e.g., televoting, games, gambling) could be displaced by SMS to normal numbers (plus access by the ASP to the account). Or worse, the SMS service itself could entirely disappear and be replaced by IP-based communication means such as e-mail, instant messaging or Web browsing.

Simpay(.org), an initiative of 4 large mobile service providers to open the mobile users' purse to third parties, probably failed when these providers realized they would open the door to competing communication services that wouldn't use the mobile infrastructure anymore, or at least not in the same way (maybe still as bitpipes).

MMS is already being offered for free* by independent MMS Centers out there on the wide open Internet, in the hope that users of these MMSCs will start using other services such as blogging, online albums, photo printing, or T-shirt printing.

Offering a stand-alone payment method could create a virtuous circle of higher revenue and lower margins, but open access to accounts has become Pandora's Box for many service providers — nobody dares to open it.

In the Internet† and credit card space, various players are trying to convince end users to establish a prepaid e-wallet, to be topped up using a credit card, or using funds transferred from another e-wallet.

Dedicated payment service providers (PSP) now allow application users to choose from a wide palette of different payment methods.

Besides within telecom operators and Internet service providers, we have pre- or postpaid accounts at banks, credit card companies, public utility companies, leasing companies, cable TV distributors, electronic purses on plastic cards, and so on. In the same line of thoughts as for the personal contact lists and tribe memberships (Section 1.9), we will soon need shared and compatible repositories for all these accounts. It is unlikely that the communication service providers will play this federating role.

There's another reason for communication service providers to be skeptical about open charging interfaces for M-commerce activities.

As communication service providers cannot complete the communications to all destinations themselves, intermediate carriers will be involved, and there will always be the possibility for these third parties to use the communication means itself as a payment method for an application. In many cases, there will be no need for a dedicated charging interface (i.e., to the BSS), separate from the communications interface (i.e., to the I-BCF or even P-CSCF).

An example of this is the use of SkypeOut credit, which is normally used to set up calls to non-Skype destinations. Upon entrance to a public parking lot, a customer could key in his Skype login credentials (login, password), and the merchant (the parking company in this case) would use the Skype API to set up a virtual call from

---

* Well, for the price of the connectivity (the Megabits of mobile data).
† For example, Paypal, eBay's online payment method.

the customer's Skype ID to a mobile number owned by a payment service provider (PSP), for example, for 10 minutes in order to pay for every hour of parking. The PSP would receive interconnect revenue, and kick back a number of cents (e.g., 5 cents a minute) to the merchant (the parking company).

A similar scheme could be deployed by NGN service providers; the destination PSP would be interconnected to the NGNSP via a SIP interface. A new technical interface (protocol standard) is thus not required. The parking lot example shows that charging to certain destinations should be units-based (e.g., sip:50eurocents@psp.ngvas.com) rather than time- or volume-based.

NGNSP could (should!) define higher payout rates for sessions to these PSP destinations. This matter is also linked to the question whether, in general, application service providers should be treated as end users or peer NGNs (Section 4.8).

Economically, on a charging interface, PSP would expect access to the NGNSP customer's purse at competitive fees, comparable to credit card fees (below 5% cost to the merchant). For a purchase amount of 100, this means a pay-out rate of 97 from the NGNSP to the PSP, followed by a kick-back of 95 from the PSP to the merchant.

The use of communication means as payment methods is not a new phenomenon in Next Generation Networks. Many Internet sites are now relying on premium calls and SMS (well-established communication means) as a charging method rather than as the content delivery channel. Users are invited to call premium numbers, send and/or receive premium SMS in order to obtain access to the site.

The vulnerability comes from malicious applications that would, using the customer's identification credentials (login and password), consume the communications credit without the customer knowing. In the era of dial-up Internet access, dialers would use the computer's modem to dial toll numbers. This threat became more stringent in NGN, where a dialer could force the local user agent to make even more invisible outbound calls over IP interfaces. Is it the NGNSP's duty to protect its end users from these security risks?

It would be of no use to develop a new, dedicated, highly secure charging API if the communications API can be broken. Security needs to be a concern on all interfaces, and the instruments to realize this are firewalls and session border control (Section 4.7).

5.13 Conclusion

For new communication means, it is of course a difficult exercise to evaluate the charging requirements in general.

The requirements could be different for retail mode vs. wholesale charging, or toward other revenue sharing parties. But for sure, huge cost savings can arise from hosting these processes in a single environment (product catalog, charging, rating logic, etc.).

A next generation communication service could certainly be launched on the basis of pure functionality only, with Skype as school example — well, at least before SkypeOut was added.

However, offering a new communication service in a sustainable fashion will require the appropriate dose of attention for each of the aspects of charging and rating. A reasonable shopping list could include the following:

> A single, solid product catalog, well balanced between a limited set of product types and an extensive set of individual options
>
> An account which is the sum of a user profile and a wallet, with a wallet usage policy allowing a single user to tap into multiple wallets, and shared wallets among corporations, families and the tribes of Section 1.11.2
>
> In general, online charging whenever possible, off-line charging as second choice
>
> Limited need for rating, by using nonmonetary credit in the wallets (e.g., seconds, megabits, clicks)
>
> Highly granular metering, to be preferred over blind openness to third party applications
>
> Reverse charging where necessary and fair to the service user
>
> Clear, real-time Advice of Charge at least toward end users
>
> Redirection to top-up and bill payment pages
>
> A more flexible payments policy, allowing to use multiple payment channels independently of the pre- or postpaid nature of an account (anyway, a single wallet could contain both pre- and postpaid credit)
>
> A well-thought promotions and loyalty strategy, after the initial growth phase

Chapter 6

Technical Requirements for a Modern VAS Platform

Along the years, communication service providers and their suppliers have accumulated quite some valuable experience to develop, test, operate, supervise and maintain Value-Added Services (VAS).

Independently of these VAS, they found some common basic principles and recurring requirements, some of which (e.g., high availability, scalability) are also applicable to the NGN core network itself, and even to certain access network elements.

Wherever it is delivered (e.g., in an access network, core network, VAS platform, billing system) a "Value-Added Service" can best be considered as an "application" surrounded by "enablers": besides the solid layered VAS platform itself (hardware servers, software architecture), also surrounding enablers such as training, well-documented operational procedures, data warehousing, financial supervision and targeted marketing, which all have great effects on the quality and success of any communication service.

Some Service Providers would therefore consider a VAS platform to also include these surrounding enablers, but here, let's limit ourselves to the technical aspects of the VAS hardware and software itself.

Let's also try and keep this chapter independent of the communication means we are going to adopt. As such, calls, messages, data sessions, multimedia sessions, and so forth, can be referred to as "communication sessions" including several "transactions."

In fact, an NGN service provider could add the VAS platform inside the user device (on top of a NGN User Agent), on servers in the core network (the most easily manageable and commonly used model) or even let "third parties" deliver the VAS (the model of application service providers). Similar requirements would arise in the three cases, for the VAS platform.

For a long time the domain of Requests for Information (RfI), Requests for Quotation (RfQ) and Requests for Proposal (RfP) by fixed and mobile communication service providers, here's an attempt to summarize the common requirements for these VAS platforms, independently of the application(s) that would be hosted on these systems.

6.1   Toward a Layered Platform Architecture

Traditional VAS platforms (prepaid systems, voicemail, etc.) used to each have their own hardware, operating system, database, data structures, service logic, "southbound" network interfaces and "northbound" interfaces for provisioning, billing and supervision. A single subscriber would have been defined in each of these platforms.

A benefit of these vertical "VAS silos" has been that they could be selected from different vendors, without too much integration effort by the communication service provider.

However, new efficiencies could be found if the VAS platforms were organized as follows (from top to bottom):

> An IT sublayer, essentially a near-real-time (off-line) environment providing the so-called "northbound interfaces" (provisioning, billing, supervision) to external IT systems, and offering a service creation environment
>
> An Applications sublayer, selecting and executing service logic based on subscriber and service data in real time (online) — not only for charging purposes but also to deliver meaningful communication services
>
> A Service Capabilities sublayer, stateful* "connectors" isolating network and protocol differences from the applications; some connectors are providing the server side to the outside world, whereas others should act as the client to third party databases; but in principle, the Service Capabilities themselves should not require access to subscriber data[†] (only applications do)
>
> An Infrastructure sublayer, consisting of best-of-breed servers, operating systems, database, protocol stacks, switches, routers, gateways for signaling and media

---

* Stateful means that such connector (or Service Capability) should maintain a communication session state (Context information), contrary to a protocol stack which only maintains a transaction state.

[†] The problem would then arise how one Service Capability would access subscriber data held on third party databases, through another Service Capability.

Ideally, a new application could automatically use all existing service capabilities, and conversely, all applications could easily be improved to make use of a new service capability. Meaning that the common "language" (e.g., API, protocol, procedures, message bus) through which the applications communicate with the connectors should be universal from start: standard but extensible, extensible but fast to decode and encode, and giving both parties a good grip on the state of the other one. It is not an evident choice.

The purpose of this section is not to compare multiple competing technologies for this language (e.g., JAIN SLEE, Parlay, Parlay X, OMA). Some of these standards are discussed in Section 3.7.

SIP itself is of course a candidate. Today's Media Servers (for packetized voice and video), for example, can already be invited in the session by back-end applications, using the SIP INVITE method. The Media Server then uses HTTP GET in order to obtain VoiceXML scripts from the back end and play them (or other media information in IETF MSML/MOML format, PacketCable NCS, etc.). SIP itself is thus not enough, and needs to be complemented or extended.

Even for quite basic call or message control, can the legacy and the new network protocols* reliably be mapped to internal SIP, by the front-end service capabilities? Will SIP allow preventing the network-specific aspects from impacting the back-end applications?

There has been a long debate on whether the "front-end" service capabilities should reside on the same physical servers as the "back-end" applications, or whether the servers should be dedicated (Service Capability Servers and Application Servers). In principle, it sounds like a waste of computing and communication resources (server I/O,† network bandwidth) to create a front-end-to-back-end dialogue that would become (due to the protocol encapsulation required) almost as consuming as the network-to-front-end communication. Moreover, inside a front-end Service Capability, the mapping of protocol states, messages, parameters and values from network side to application side should remain a vendor-specific competence. Standardization attempts have failed. Let's further assume that the VAS AS run both the applications and the service capabilities.

As a quite unrelated topic, service creation is often only possible in the Application layer, that is, through graphical environments where service logic can be modified.

Few service creation environments exist which allow editing database object structures, and automatically adapting the IT aspects such as screens, security permissions, provisioning commands or call tickets. Therefore, developing the IT aspects for a new communications application today often amounts to 90% of the effort to develop the service.

---

* ITU-T INAP, ISUP, H.323, 3GPP CAMEL, MAP, IETF RADIUS, DIAMETER, etc.
† Input/Output.

Different service creation environments must be used to create new connectors in the Service Capabilities layer, for example tools based on the Object Management Group's (OMG) Unified Modeling Language (UML).

In the Infrastructure layer, the biggest effort is probably to continuously requalify and adapt the software (service capabilities and applications) to an ever evolving environment of servers, operating systems, databases and gateways. The Advanced Telecommunications Computing Architecture (AdvancedTCA) using Carrier Grade Linux is today the most important standardization effort in the server hardware area.

Layering is of course a popular concept and has been used to justify "convergent billing" (i.e., in fact generalized online charging) or the NGN architecture itself (see Section 3.1).

Layering cannot become a goal in itself. Establishing the communication between the layers is a long term investment; in the short term, the cost and complexity of adding another VAS silo will generally be lower and the performance (the handling capacity of communication events) will be higher. A lot of processing power can indeed be spent on inter-layer communication rather than on actual network interaction.

The layered architecture of these platforms should ideally be reflected in matching organizations, on telecom equipment manufacturer/software vendor side as well as on telecom service provider side. Otherwise, even the most perfectly layered technical platform will not lead to cost savings nor noticeable service quality improvements.

6.2 Load Balancing

VAS platforms sometimes initiate transactions to core networks, but for most of the time respond to requests from these networks.

In SS7 networks, SCCP Relay Nodes are able to spread or balance the load of messages across a set of Signaling End Points (e.g., SS7-to-SIGTRAN signaling gateways). The MTP layer underneath then ensures that multiple routes are available to reach that Signaling End Point.

SIGTRAN network elements (e.g., the signaling gateway mentioned above) use an Address Mapping Function to share the load across a set of SIGTRAN hosts.

Content Service Switches have been developed to balance a load of (Layer 7) HTTP requests across a Web server farm (Layer 4). These CSS are today also selecting the target Web server farm using advanced criteria such as proximity, user prioritization, client device or requested content type. Smart functions have been added, for example, to off-load SSL processing (Layer 5) from the Web servers, hence accelerate content delivery.

Similar network solutions now exist for SIP.

The call routing logic of telephony exchanges (fixed and mobile) is almost always designed in three stages: analysis objects to (pre)analyze the called party number, routing objects a.o. to spread the load across multiple destinations, and trunk groups (groups of circuits) to reach that next destination.

In general, effective load balancing is only possible using timely information about the target server's state, and/or the communication application itself on that server (if the server itself is reachable).

For example, by default, SIP (user agents or proxy servers) and HTTP clients use Domain Name Service (DNS) resolution to direct the SIP messages to the destination servers (e.g., SIP registrars, proxies or Application Servers). The DNS could return various answers in order to spread the load of SIP requests (this is known as the DNS round robin mode). However, the DNS servers are not informed of sudden unavailability of a destination SIP or HTTP server. Hence they cannot react to such event, and the client needs to discover such unavailability itself, through time-out (e.g., 30 seconds). It then needs to use a back-up server, indicated in the DNS response.

So when designing a VAS platform, we should thus first think about a watchdog process (see Section 6.10) that will monitor the applications not only to restart them in case of failure, but also to inform the third party load balancers.

Second, if the VAS platform is going to be used for the entire duration of the communication, the load balancer should also maintain (count) the number of communications in progress to each VAS platform and application.

Third, within a VAS platform itself, which can consist of multiple geographically remote physical servers, each layer (of Section 6.1) should be designed to share the load across the layer above it. This means that each layer should offer a degree of redundancy: the ability for separate modules to execute the same function. We'll address that in Section 6.3.

In the long term, one can wonder whether both network-based load sharing and platform-internal load sharing are both required. Let's say here that the former should rely on open standards, whereas the latter can be optimally designed within the VAS platforms, hence be more accurate. Designing too advanced internal load balancing schemes may end up consuming an exaggerated amount of computing and memory resources, compared to the service execution itself.

6.3 Redundancy

As introduced above, redundancy implies that multiple modules should be able to deliver the same service.

In the IT layer, multiple servers can be used to hold the platform's clustered master database; each server is connected to the shared disk array. Solutions exist to back-up the database to a remote site (server and disk array) periodically, in near real-time, should a catastrophic outage occur in the main disk array.

On the real-time Application Servers, multiple instances (independent UNIX processes) can be used to avoid that the failure of one process would bring the whole AS down. Each so-called back-end process should reserve its own portion of memory to store the Context information (e.g., the parameters of the calls, messages, sessions in progress, subscriber and service data, counters, timers).

In the Service Capabilities layer, multiple instances can also be used to implement a single "front-end" service capability (such as a messaging interface), but usually they end up listening to a particular communication port (TCP, UDP port number) or pool of port numbers of the server's local IP stack. Hence, load balancing across the service capability instances of a single server is normally network-based. Of course, one such service capability instance should be able to balance the load of requests to multiple application instances, based on their current load and availability.

At Infrastructure level, mirrored disks should protect the data from corruption. Multiple physical (Ethernet) interfaces should connect the server to a single external LAN switch. And a single server should be connected to multiple LAN switches. That single server should be able to use multiple IP routers (behind these LAN switches) to reach any IP destination. Routing protocols should be appropriately configured to inform the local servers of the availability of particular routes (i.e., next hops) to reach any particular destination — and, conversely, to inform the external world of the IP reachability of the local servers.

After having performed load balancing, it is important for external load balancers to direct all subsequent requests (messages, packets) of a same communication session, to the same physical server (service capability instance). In the network, a frequently used technique is to assign a general network address (e.g., SCCP Global Title, HTTP/ SIP URI, host name) to an application, but then to let each server or service capability instance indicate its own network address in the first response to the external client.

Is this also the case for a service capability instance toward the set of application instances? In principle, yes, all subsequent requests should go to the same application instance: only a given application instance has access to the necessary context data to handle the current communication session.

This means that, if the application instance fails in the busy hour, all sessions handled by that instance will be lost. In practice that can correspond to, for example, 100K€ of prepaid credit, allocated to ongoing communication sessions, and which cannot be debited from the accounts.

Whereas this sounds unacceptable for a banking transactions server, most communications service providers accept this risk (of loss of context data).

But in more advanced designs, the application instances can share common functions (e.g., running timers, incrementing counters) to a dedicated process, which again can exist in multiple instances.

Next, we'll discuss database replication in Section 6.4, and memory synchronization in 6.5, as techniques to share the information among multiple application instances.

6.4 Database Replication

All layers, except the Infrastructure layer itself, require data (e.g., subscribers, service configuration data, CDRs, alarms, statistics) to be stored persistently for days, if not weeks or years. It is unacceptable to keep these data in memory only, and assume that the servers will never fail or need to be upgraded.

In this context, the pace at which these data change is referred to as the database volatility.

As the added value of a service increases, so does database volatility.

Data can be frequently altered at provisioning time, through user self-care, but increasingly due to network events. Communication service providers expect good visibility on the communications sessions, the attempts to set up such sessions, the current health state of the platform and all activities executed (see Section 6.26).

The IT layers of VAS platforms are sometimes evolving into real "data warehouses" for marketing studies on user behavior.

Persistent data should be propagated from the IT layer's master database into the real-time databases on the AS. The application and service capability instances should read from their *local* real-time database* *only*.

When one AS needs to alter the persistent data, it could request the update to one peer AS, to all peer AS, or to the IT layer. From the IT layer, replication to the other AS can be more selective and configurable (to get only the correct pieces of information in the correct AS).

In peer-to-peer networks, replications happens to a subset of peer AS: a set of "super nodes" maintaining the persistent data. Try installing the Skype client on a different computer and logging in with your credentials: it takes a while for Skype to replicate your list of contacts from one of these super nodes into your new computer.

Whatever strategy is adopted, pending database replication requests (from AS to IT or vice versa) should be cached in memory or written to the local database, should the connectivity disappear or should the remote AS be out of service. As long as there's available space allocated in the replication buffers, automatic database resynchronization should be possible when the connectivity is restored.

In case of longer outages, manual database resynchronization is possible.

Database replication schemes should never be used for communication context data. The consequence of writing these data to disk would dramatically handicap the execution of service logic.

6.5 Memory Synchronization

Horizontal memory synchronization is the only viable method to propagate context information across the AS farm, hence to reduce the risks associated with the loss of context data.

Usually it will be performed to the peer AS (of an active/standby pair); but why not to several AS, if the context data can be held in a sufficiently compact format.

Also here, buffering and resynchronization after outages is a must.

Compact, binary data structures are highly beneficial to database replication and memory synchronization, but other factors come into play when selecting a database schema (see Section 6.20).

---

* As mentioned, applications may also read from third party databases, through certain service capabilities.

6.6 High Availability

High Availability (HA) is the result of technical but also operational factors.

Most VAS and NGN platform vendors claim 99,999% availability, in general.

But communication service providers have come to rely on more precise Service Level Agreements (SLAs) with their suppliers.

In these contracts, it has been very difficult to come to a clear definition of service availability, as the communication service consists of the application and all surrounding enablers (IT, Service Capabilities and Infrastructure).

Also, the distinction between congestion, partial unavailability, and outage, or the classification of faults into severity levels (i.e., critical, major, minor), has been a point of lengthy discussions.

A high degree of redundancy alone does not ensure HA. For example, if all ongoing communications are lost when an application instance fails, the platform can still be considered to be redundant (if there are other application instances), but it is obviously less HA than if there would be memory synchronization for the context information. And even then, a memory leak (an application reserving more memory than it deallocates for each context) could occur in a perfectly redundant system, probably affecting all application instances at once.

6.7 Scalability

Another popular confusion is between redundancy and scalability.

Very redundant systems can turn out to be totally nonscalable.

When adding the next (often more powerful) AS, database replication, memory synchronization or other processes (CDR collection) can turn out to burn an ever increasing amount of computing and communication resources, until all previous AS have to be replaced by more powerful models.

Functional scalability is also to be considered: it cannot be the case that adding online charging, for example, totally ruins the performance of a number translation service.

It is useful to have, from start, a dimensioning (simulation) tool which predicts resource utilization (CPU, memory, database, I/O) for both functional and capacity upgrades, along the years.

6.8 Resource Allocation

Resource utilization is of course strongly depending on the resource allocation policy.

A multitasking OS can make sure it grants sufficient CPU power and I/O access to different processes, also based on their configurable relative priority.

A database can also be tuned to provide optimal response times to the most critical users (applications).

But memory allocation is typically not governed by the infrastructure layer.

It should be possible for protocol stacks and service capabilities to receive invalid, nonsupported or forbidden requests from the network, without memory being needlessly reserved to hold a transaction or communication state.

If the network request is valid and permitted, Service Capabilities, by definition, rely on resource allocation to store the context information pertaining to the communication session.

Applications must be able to extend this reservation, for example, to append other transactions to the communication session.

But before the Application intervenes, Service selection uses the context information, as mentioned in Section 6.9 below.

6.9 Service Selection

When a communication service request (e.g., SIP INVITE) comes in from the network, it is important to design a good strategy to map the parameters of the request to the correct application instance.

But often, the actual application service logic to be applied cannot be determined before the subscriber and service database has been consulted (e.g., to discover the originating user's product type).

This process of service selection too often ends up being performed at the start of the execution of a "default" application. When a new application is added, the network or service capabilities still select the default application, and application chaining is then required (to pass service control from one application to the next one). Introducing a new service ends up impacting all existing default applications on the AS.

The network and service capabilities should thus invoke the service selection process first, by default.

Application instances should be able to signal their state and load to the service selection process, for it to take action. It is therefore not a good idea to build the service selection process into each service capability: they would all need to be informed.

The service selection problem is different from service chaining: the service selection process should normally select only one application to be executed.

6.10 Watchdog

It would be naive to assume that service capability instances, service selection instances, application instances, and common processes will never become congested or fail.

A highly available platform should therefore include a watchdog process, continuously polling the other processes with "are you OK?" and "are you congested?" messages, and taking appropriate action upon time-out or negative responses.

The watchdog process should restart failed processes and instruct the service selection instances to divert new requests to non-overloaded application instances.

6.11 Overload Protection

Overload protection is the ability of the VAS platform to either reject incoming requests from (the Session layer of) the network, or to instruct to throttle (lower) the number of requests.

Whatever is used (rejection or throttling), the VAS platform should maintain counters of network events per second, per application, but perhaps also per origin, destination, and so forth, if communication requests should be rejected at that more granular level.

The overload thresholds (limits as of which overload is detected) may either be preprovisioned (static mode) or the system may, in real time, discover a top-10 of applications, origins and destinations causing the overload (dynamic mode).

The counters and thresholds may be applicable within each VAS server (platform-specific), or maintained and enforced across the full set of VAS servers (application-wide).

Platform-specific overload control means that if one server is overloaded, the network may decide to route new requests to the remaining other servers, which are usually also busy at that moment, therefore causing a chain reaction. Therefore the platform-specific approach to overload control is forbidden in cases where the core network elements cannot be instructed to lower the number of requests (rather than redistributing them to different VAS servers).

The application-wide overload protection approach requires a smart way to communicate counters and overload events between the VAS servers, as otherwise the VAS severs might end up consuming more resources *talking about the overload* than actually *resolving it.*

Whichever of these two methods is used, if the VAS platform simply rejects the network's request because of overload, the network should of course not simply retry that request to the same server. Redistributing it to a different one is as pernicious.

The originating (calling or message sending) user needs to be informed promptly and properly, otherwise he/she will simply retry, not knowing the nature of the problem.

If the network can be instructed (by the VAS platforms) to throttle (lower) the number of requests, this method should be preferred, but a number of extra issues is to be addressed.

What level of granularity is supported in the network (e.g., by application, origin, destination)? The application to be invoked, for example, is sometimes unknown by the network — only the fact that *an application* needs to be invoked for a given subscribe, number range or domain.

Second, it should be avoided that a network element receives conflicting instructions from multiple VAS servers, therefore a pair of servers (active + stand-by) must be defined (configured) to be responsible for each application, origin and destination. The responsible pair collects the counters from the other servers and takes the decision to ask all the network elements for throttling.

Third, the conditions should be defined in which the throttling will be stopped (or at least, reduced). If this happens too fast, the rate at which throttling instructions are given by the VAS layer will be too high.

Any overload protection strategy is strongly dependent on the network and user device to accept instructions to retry later, indicating the reason (temporary overload).

Overload can also occur in different processes of the VAS platform — not only in service execution. Database access is an example. Batch operations on the database (e.g., to provision new subscribers, clean up old CDRs) can deprive the application execution processes.

6.12 Software Distribution

The software on a complex VAS platform will evolve over time through multiple versions, under influence of new requirements, but also in order to correct errors discovered by operators worldwide.

The layered architecture described in Section 6.1 is not a guarantee that changes within one layer will not impact components in the other layers. For example, new application features often go hand in hand with new database objects (in the Infrastructure layer), and screens, CDRs or subscriber provisioning commands (in the IT layer). Service Capabilities are usually more isolated, having less impact on the other layers but still, they also produce CDRs, alarms, statistics, and so on, which need to be processed in the IT layer.

A convenient way to deal with this is for the VAS platform provider to supply software "packages," containing the multiple components required to introduce a given functionality.

Packages should also contain the install process used to install them; sometimes this is an interactive process where the Operator needs to fill in license keys, configuration preferences, and so on. Also the uninstall process should be part of any package.

The package install process may need to check the presence of other packages before attempting to remove a component, if it may be shared by components of different packages.

Even updates to documentation or training materials can be part of a package.

At any time it should be possible to check the full list of installed packages on the servers, preferably from a central point.

These are all very common concepts in IT industry, but VAS providers have not always offered the appropriate level of rigor, transparency and consistency.

6.13   Software Versioning

At each layer of the architecture of Section 6.1, the version will have a different significance; examples are a different screen layout in the IT layer, a more refined service logic in the Applications layer, support for new network interfaces in the Service Capabilities layer, or a new database object model in the Infrastructure layer.

Of course within a single layer, multiple software components will need to coexist, each dedicated to a single task and therefore having its own lifecycle of improvements and corrections.

Many problems can be avoided with an upfront, clear and unambiguous definition of a software versioning scheme.

Version numbers should be given to software components, *not to packages*. The package should therefore clearly indicate the version number of each component.

Four digits are usually sufficient to qualify the version of a software component, in the format X.Y.Z.P where:

> X is best used to indicate a Major Upgrade of the component, *affecting the way existing functionality is executed*, together with introducing new functionality.
>
> Y should denote a Minor Upgrade, introducing significant new functionality *but not affecting the way existing functionality is performed.*
>
> Z should denote an Update, introducing limited new functionality that may even be specific to a group of customers (as opposed to Minor Upgrades which should be kept customer-independent!).
>
> P should denote a Patch level, not introducing any new functionality and only correcting bugs of different severity (of course, Updates, Minor and Major Upgrades should include fixes for all known bugs at their time of issue).

This definition is only a general canvas, from experience, but may in practice further be refined by suppliers and operators.

The purpose of this definition is not only clarity during operational discussions, but also in commercial, contractual and legal domains. For example, it is very important in a support contract to specify which software versions will be supplied and/or installed for free by the supplier; which ones will be supplied/installed subject to a fixed agreed price, and which ones will be subject to new, independent commercial offers.

6.14   Software Release Cycle

Each software version goes through different phases in the software production process — from concept, over scoping and planning, requirements, high-level design, detailed design, unit testing, internal functional testing including nonregression, compatibility and performance, to early field trial, general availability, and so on. The know-how and experience of the software vendor are key to evaluate which steps to perform for which type of component and version.

It is possible that not all intermediate versions (especially Patch levels) should be installed in order to update or upgrade a platform to the latest available version.

Also there needs to be a well-documented procedure to announce the end of life (EOL) of a version or the component itself, and the consequences in terms of vendor support: usually only Patches are available for severe bugs and for a limited time after the EOL announcement.

The release cycle greatly influences the total cost of ownership (TCO).

Low-quality software will result in many patches (for severe bugs), updates, upgrades, forcing the operator to engage significant human, capital and operational expenditure even to keep a comparable level of functionality throughout the years.

High-quality software will consist of major upgrades (e.g., one every two years) that contain enough functionality, in advance of a low number of minor upgrades (e.g., two a year). Operator-specific updates will be delivered/installed for free and without much intervention from the operator. There will be almost no need to install patches; the severity of bugs will be low and the patches could wait for the next update.

6.15 Backward Compatibility

Each new version of a component brings along the risk to impact surrounding components.

If these components are all within a same package, usually the versions in that package will have been tested by the vendor to interoperate correctly.

Things get fuzzier when packages introduce new components (or versions) which need to interoperate with different existing versions of other components (i.e., outside of the new package).

Therefore it is a good practice by the vendor to maintain and publish a software compatibility grid or workbook. For all versions of any component, the compatibility grid should indicate with which other components (and their version) backward compatibility is

Ensured and verified
Assumed but nontested
Not assumed but nontested
Tested, resulting in noncompatibility

6.16 Version Activation

It is an operational nightmare having to install new software packages and components at night in the so-called "operational window," right before they go into production in the morning.

Ideally, packages should be installed well in advance, without removing the current components and versions being used in production.

The VAS platform would then include a version management and activation process allowing traffic (and screens, provisioning commands, etc) to be switched from one version of a component to a next one. In fact, allowing the components of an entire package to exist in multiple versions, installed simultaneously on the same platform.

To avoid excessive backwards compatibility requirements, all components of a single package should be brought into production at once, but this is sometimes not easy. Imagine the simultaneous traffic, subscriber data, and provisioning screens migration during a major upgrade. Should a new version of the subscriber management screens (or provisioning commands) be made compatible with the previous version of the database structures?

The support of multiple versions running in parallel on a VAS platform remains, in practice, a noble objective to achieve. One with major benefits in terms of operational expenses, but only to be realized for software patches and updates (not for minor and major upgrades).

6.17 Licensing

Software licensing is an extensive area covering technical aspects (e.g., what is allowed to be used, how frequently, how to measure) but also commercial, contractual and legal aspects.

Is a license granted for a package, for a software product (consisting of multiple packages), or for a solution (a combination of products)?

What happens, for example, when an operator using the software license is acquired by a parent company also using the same product, but in a different version? Are the two licenses somehow merged?

In this little section let's focus on the technical aspects, assuming that the license is granted for a software product with a feature road map.

The basic question is what metric will be used to identify the usage of the software product:

> Site-wide license: this unlimited usage will usually be affordable by very large corporate organizations.
> Duration for which the license is required: a free test or evaluation license may be granted for a limited period.
> Number of CPUs of the servers on which the product is installed: unfair if only some components are installed on some servers (front end Service Capability servers) and other components on different ones (back end Application Servers). Also unfair if the software is deployed on N+M servers for redundancy. And this scheme doesn't offer an incentive to the

software vendor to start supporting the latest (more powerful) CPUs, as
the overall license revenue will simply be reduced.

Number of telecom events being handled by the platform: successful only
or also unsuccessful communication attempts?

Number of potential (i.e., usually preprovisioned) users of the VAS: active
only, or also users which have obviously stopped using the service but are
still provisioned on the system?

Application-specific metrics such as number of scratch vouchers, credit top-
up operations, voice messages, presence notifications.

Any traffic-related metric will have to be very accurately defined as a peak value,
an average or a median value, over a given measurement period (e.g., 1 hour), with a
number of measurements (e.g., 3600) and granularity (e.g., 1 second).

License measurement could be a continuous process, whereby the last N measurement periods are averaged out (or the peak or median is calculated).

Alternatively, the time-of-week and day-of-year should be specified on which
such measurements (a period of 1 hour in the example) will be conducted; maybe
it's worth not taking into account exceptional events such as New Year's eve for
commercial license calculation.

Above a given usage threshold, the VAS system may even automatically stop
accepting new requests, but we think this shall be done only to protect the servers from crashing (overload protection), not to enforce a commercial license. As
a golden design and operational rule, the infrastructural capacity should always
exceed the commercially licensed capacity.

6.18   Openness

We already discussed the potential need for openness of the prepaid or postpaid
subscriber account in Section 5.12.

Different forms of "openness" will also be required on the VAS platform:

Of course, the openness of the provisioning, billing and supervision interfaces to the northbound IT systems (OSS, BSS and TMN).

The ability for third-party applications to use the Service Capabilities of
the platform, via open Application Programming Interfaces, and vice
versa; this openness should not be over-valued by communication service
providers, as there are almost no such "third parties" able to develop components for multiple systems.

The ability for the VAS platform's IT layer (local supervision screens) to
incorporate alarms, statistics and configuration of closely linked third
party network elements such as Signaling GateWays, Media Servers or
Media GateWays.

6.19   Modularity

Software is modular when the functionality is broken down to appropriately sized, nonoverlapping entities. These entities are typically one notch lower in the hierarchy than the components we discussed in Section 6.12.

Using a reasonable number of Dynamic Linked Libraries (.dll), each dedicated to an understandable function, for example, decoding/encoding a given protocol, is a good example of modular design.

An average Service Capability, for example, would need to speak at least a southbound protocol (toward the network) and a northbound protocol (toward the application), plus libraries for TMN FCAPS functions and so forth.

Hence the rules outlined in Sections 6.1 and 6.12 unfortunately do not prevent telecom software (VAS platform software in this case) to become nonmodular. A same protocol library is included as different code in two service capabilities; different applications start using different code for database access or CDR generation, different screen panels and underlying database tables include similar subscriber information, and so on.

As this level of detail is usually visible only in the detailed design documents and in the source code itself, it is difficult for the communication service provider to assess the modularity of the code.

6.20   Data Model

The data model is a representation of the persistent information in the subscriber and service database. It can also be used to describe in-memory information such as call contexts or protocol state machines.

Usually a call model comes as a diagram of database tables, linked to each other via one-to-many and many-to-many relationships.

It helps people understand the existing information, and to imagine how new information can best be added.

A data model needs to be modular, avoiding tables with too many columns or too few rows.

There should only be one data model for the entire VAS platform, not for each Application. Transgressing this rule invariably results in silo applications, where the object "Subscriber" exists within each application silo, in a slightly different form.

Having a single data model forces the VAS application developer to recognize the fact that the subscriber may have subscribed to multiple applications at once.

A good data model puts the Subscriber at the center. But, often, large groups of subscribers will share the same values for most of the attributes. Repeating this information in thousands or millions of rows is an inefficient approach. Therefore, the common properties of a group of Subscribers should be defined as a Product. This Product may be application-specific or platform-wide.

However, with today's highly personalized VAS, there's a risk of explosion of the number of Products. It is indeed tempting, whenever a new feature X needs to be introduced, to break down Product A into (i.e., divide Subscribers among) a Product A1 containing the new feature and a Product A2 not containing it. The next feature comes along, and Products A11, A12, A21, and A22 appear.

Therefore it is interesting and necessary to create the object Option too, linked to the Subscriber in a one-to-many relationship, and to the Product.

The communication service provider will acquire subscribers through different direct and indirect channels, with potentially different requirements in terms of available applications and features. It is a good idea to link the Subscribers and the Product Types to a Channel object. Different Channels could also represent internal and test subscribers. CRM agents belonging to a given Channel should be entitled to view and update only their "slice" (Subscribers, Products) of the system.

The platform-wide Subscriber-Product-Option-Channel paradigm has proven to be very difficult to add, once different Subscriber objects had been defined in multiple applications.

It is therefore something to be enforced at the heart of the VAS platform, as part of the fundamental infrastructure, or even throughout the entire communication service provider organization.

6.21 Storage Strategy

The persistent data model needs to be stored in nonvolatile media: typically, hard disk drives.

These can be located inside each VAS server, or centrally deployed as disk arrays,* accessed by the servers via a SAN (Storage Access Network) — usually using optical fiber channels. Hard disk controllers (whether inside the server, or in central disk arrays), will normally write the same information on two hard disks, a principle called mirroring.

Whereas mirroring is a must, the degree of redundancy of the disk arrays and of the (single or dual) controllers inside each disk array is to be appreciated by the VAS supplier and the NGN communication service provider.

The inconvenient truth about SAN is that they require immense and unpredictable bandwidth between a server on site A and a disk array on site B (in the order of a few to hundreds of Mbit/s), which is a problem in wide area and transnational set-ups. The bandwidth needs to be purchased from transmission providers (e.g., STM-1, OC-3) rather than from Internet service providers — a much more competitive sector.

Therefore the new trend in IT industry is called Virtualization: database servers on central sites are accessed by decentralized, remotely located application servers (VAS servers in this case), using SQL*Net, LDAP or other protocols; over MPLS/IP

---

* Not to be confused with a database server.

networks, usually with end-to-end Security and Quality-of-Service enforcement. The throughput requirements are however not much lower than in SAN, even though previous responses can be cached in the application servers, if the data are not too volatile. Also, only persistent database information can be shared this way (no transient in-memory data). Finally, Virtualization is not a guarantee that the actual data will be stored on multiple sites. For a disaster-proof architecture, it is up to the application developer to make sure (or preferably at VAS Infrastructure level) that all processes write to two different databases (Data Source Names) located on different sites.

There's a third possible storage strategy, besides SAN (brutal force) and Virtualization (a partial solution). But it requires the VAS platform supplier to develop his own selective Database Replication strategy (top-down — cf. Section 6.4) and Memory Synchronization (horizontal — cf. Section 6.5). The in-memory data will namely often end up being written and committed to the database too. This approach will be less bandwidth-hungry, as storage is local (inside the servers or disk arrays on a local SAN). Only data requiring to be replicated (insert, update and delete queries), are actually transmitted to the central master replication servers or peer servers. Read queries are performed locally.

We believe it is a good idea to align the storage strategy for both the Application and IT layers of the VAS platforms. Today, with full throttle on Virtualization of IT infrastructures, but little attention to VAS platforms, this is seldom the case.

6.22 Backup and Recovery

Besides holding the same data, it is important for the Application Servers and IT servers to benefit from back-up and recovery procedures to and from the so-called disaster recovery (DR) sites.

DR is a booming business, which is, by the way, still a candidate to be outsourced from many organizations.

DR sites have become the atomic bunkers of any sustainable business (e.g., communication, banking, insurance, airlines). And it is tempting to test the evacuation procedure to these bunkers once and for all, namely, upon construction of the bunker. Nobody also worries about what will happen when people ultimately get out of such bunker, and huge data streams flow out of their DR sites.

Now speaking about the scope of the backup, everyone would certainly agree that subscriber, account, and CDR databases should be backed up. IT administrators could already be more skeptical about configuration files and application software, but what if it took 10 hours to reinstall the software packages and all patches, in order to restore a failed system?

6.23 Remote Access

Remote access to a park of VAS servers should happen through a single entry point (e.g., a terminal server) or at least a single sign-on procedure.

Indeed, it is inconceivable to define UNIX logins, passwords and file permissions on a per machine basis.

The topic of remote access brings up a related topic, namely the passwords policy. It should not be allowed to define trivial general password acceptance and refreshment policies. Trivial passwords should not be accepted; the user should be informed about the estimated strength of his newly chosen password, and access should not be permitted without the user regularly being forced to refresh his/her password.

Of course the remote access pipe needs to be secure. Various IP VPN products allow remote support personnel to access the company LAN through the public Internet, but their permissions should be limited to a given set of machines and logins.

Limiting the access to only a set of machines is a poor option, as support teams from different software vendors should be able to log in to a shared VAS application server, and see only their application's directories and files (or only the database, for example) — perhaps a noble but unachievable goal.

6.24 Hosting

The all-IP NGN and VAS platform architecture calls for hardware, OS and software hosting.

Hosting is today known as the activity to rent and maintain computer hardware, operating systems, databases, remote access servers, firewalls, rack space, reliable power, cooling, and LAN and IP connectivity for either NGN service providers or VAS software suppliers.

Hosting is providing a hotel for applications.

And in a hotel, check-in and check-out are among the more important procedures influencing customer satisfaction.

6.25 Reporting

Reporting is too often confused with performance management (which we'll address in Section 6.26).

Periodic, triggered and on-demand reports should inform the NGNSP about the uptake of communication means and value-added services, both in terms of subscription and usage.

This starts from the number of subscribers having subscribed, via a given commercial channel, to a given service, product type and individual options — the structure we discussed in Section 5.1.

Next, the usage needs to be analyzed not only in terms of successful communication events per service, product type, option and destination, but also in terms of failed session establishment attempts.

And certainly, the NGNSP should get a good view of how *infrequently* certain options are used, in order to simplify the commercial offer.

Finally, an important aspect of reporting is that it should allow a NGNSP to detect a decrease in usage by certain partners (peer NGN, ASP's) or subscribers. Today this is often considered to be a data warehouse function, generating a weekly report, but in an increasingly volatile NGN environment, a sudden decrease in traffic from a given source can already indicate that an important peer, ASP or user agent has selected another NGNSP as the next hop. Conversely, a too interesting offer can suddenly attract far too much traffic and result in Service Level Agreements not being met anymore.

It is not a subscriber decision anymore.

For all these reasons, reports cannot be generated by post-processing CDRs collected from the core network and VAS platforms.

The VAS platform itself needs to generate a configurable set of reports, a set of counters and even alarms on these counters (thresholds), informing the NGNSP's commercial chain about a corrective action to be taken (advertising a new price, for example).

This commercial battle can be won only with sufficient automation.

6.26 TMN FCAPS

The Telecommunications Management Network (TMN) is originally a protocol model defined by the International Standards Organization (ISO), then ITU-T with the M.3000 series of recommendations. TMN was structured in Business, Service, Network and Element management. It included a protocol called Common Management Information Protocol (CMIP, ITU-T X.700 series), which today is largely superseded by the IETF Simple Network Management Protocol (SNMP).

SNMP defines an Agent, located on the VAS platform in this case, communicating with an external Network Management System (NMS), which is shared by many network systems (e.g., access network, core network, VAS platforms). The Agent can eventually be split up as a Master Agent and Subagents, each responsible for a particular subsystem.

The Management Information Base (MIB) defines the objects being managed and their properties. For a VAS platform, the quality and documentation of a MIB is an indicator for the quality of supervision to be expected (does it cover the software components?).

Today, in practice TMN compatibility means that a VAS platform and the applications should be supervised or managed by 5 functions:

> Fault management: alarms of various severity levels are sent by the Agent to the NMS, as SNMP Traps. An important role of the NMS is to correlate alarms produced by different network systems, and consecutive alarms produced repeatedly, to be due to a single event.

Configuration management: the NMS should be able to modify configuration details (e.g., addresses, timers, table sizes) in the VAS platform — this is very seldom the case.

Accounting management: the VAS platform should produce a CDR for each service event, and transport it to the BSS.

Performance management: the VAS platform should produce statistics counters regarding each object in the MIB, and periodically send them to the NMS.

Security management: the VAS platform should protect itself from intrusion by nonauthorized persons, a.o. with individual logins also to application screens, a periodic passwords reset policy, etc.

6.27 Migration Tools

Subscriber and service database structures in a VAS platform will be enriched over time, as new functions are plugged in to the VAS software. A migration tool will extract the subscriber and service data used by the previous application software release, manipulated and reinsert it in the database, to be used by the next release.

This process can happen in batches of say 10,000 subscribers, or as a continuous process (one by one). But it is important to include the roll-back functionality, the ability to restore the previous situation before the batch, or even the initial situation.

A major new application software release will also bring new screens, MML provisioning commands, CDRs, alarms, statistics, reports, and so on. In most cases, it will be unfeasible to upgrade the neighboring IT systems to switch to the new version overnight, or support two versions simultaneously (e.g., old provisioning commands for nonmigrated subscribers, and new commands for the migrated and new ones).

Here, the best strategy is probably to install the new software release supporting both the old and the new version of the MML, switch these external interfaces to use the new MML one by one, migrate all subscribers (by batch or continuous process), and then only stop supporting the old MML. The same is true for screens and the other components. It sounds like common sense but how many times is only the application software per se considered to be a user of the subscriber and service data?

Migration tools should thus include "railroad switch" processes between old and new versions of a provisioning interface, screens server or alarms generating process.

During major upgrades, hardware, OS and databases will also need to be replaced. Most of these changes can however be performed in production environments, without requiring migration tools, thanks to the N + 1 or 2N redundant design of the VAS platform.

A VAS platform architecture needs to be scalable not only from the quantitative perspective but increasingly from a functional perspective.

6.28 Documentation

Here we arrive at the edge of the VAS platform features, and enter the domain of what we called the surrounding enablers in the introduction of this chapter.

Documentation should be built into provisioning, operations and supervision screens (as hyperlinks) and MML commands (as help pages) as much as possible.

The days are over when CRM agents, operations and supervision engineers could be trained for weeks on the signification of fields, on processes and available features.

A new application or platform software release should thus also come with the appropriate new documentation modules.

6.29 Support

Platform and application software support is another, yet even more important enabler of modern communication services.

It would require a dedicated book to describe the aspects of a competitive support process. The Service-Level Agreement (SLA) between service provider and platform or software supplier would be one of the chapters.

Strictly from a VAS platform perspective, the supportability of a VAS platform and software of course depends on almost all characteristics we discussed in this chapter.

Supportability, which can be expressed, for example, as a reasonable number of cases and rapid resolution times, is the reward for service providers and suppliers who have respected these principles of VAS platform design.

Chapter 7

New Generation of Value-Added Services

In the past chapters, we have made an attempt to do the following:

Characterize communication means, in Chapter 1

Examine the communication service providers' starting position, in Chapter 2

Summarize the networks proposed by the standards bodies, in Chapter 3

Interconnect these networks, in Chapter 4

Evaluate the charging requirements, in Chapter 5

Produce the wish list for the service platforms that will host the new applications, in Chapter 6

We should now have a good idea of the environment in which successful Value-Added Services (VAS) could be created, at least from a technical perspective.

Our macro-view in Section 2.5 even suggested that there could be room for new VAS proposals, to be accessed by enterprise telephones and mobile data devices:

Will basic TISPAN NGN service (establishing SIP sessions, with QoS enforcement and SBC functions) at $10 a month be considered superior to Plain Old Telephone Service (PSTN) at $6 a month?

Would you invest in your own $35-a-head, fully serviced enterprise IP PBX, or would you rather rely on a $10-a-month voice and video centrex offer from your NGN service provider?

Rather than subscribing to a $35-a-month IPTV program, digital DVB-H or even plain old analogue broadcasting at $20 a month, people might pay a $5 to $10 premium a month to view YouTube videos on large screens, in a high-definition video version — with some IP QoS enforced.

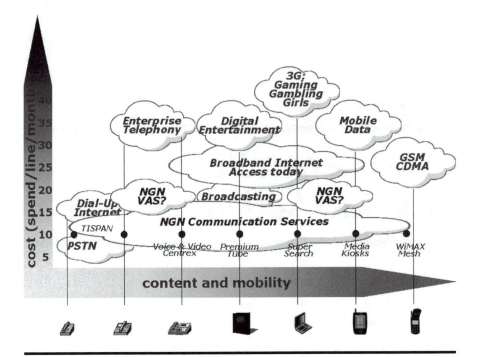

Figure 7.1 Macroeconomic perspective on the opportunities for NGN VAS.

Rather than visiting various gaming, gambling and girls sites,* and spending an unpredictable $40-a-month on heteroclite and perhaps untrustworthy sites, the less adventurous among us might spend $10 on a super search site offering only free previews, trials and contests.

Why spend $35 a month on mobile data connectivity when for $10 you could subscribe to a media kiosk at your federation of WiFi hotspots, automatically downloading your selection of electronic newspapers and magazines to your PDA, PSP, or media player, whenever you'd pass by such hotspot.

Would you prefer to spend $10 a month on quite patchy, flaky and meshed WiMAX coverage, allowing you to run your own VoIP user agent, rather than $30 a month on ubiquitous GSM + GPRS coverage forcing you to wait 1 second for a reply to ping?

At this point, most telecom service providers would argue that our macro-level and technology-driven approach needs to be counterbalanced with a marketing exercise.

This chapter would then define the marketing mix (Product, Pricing, Promotion, Placement/distribution — the four Ps) for individual VAS or bundles, in an attempt to position them to a segmented population of communication service consumers.

---

* The popular joke about 3G services.

OK, service marketing is quite different from the marketing of Fast Moving Consumer Goods, and requires an extended marketing mix adding three P's (People, Process and Physical evidence). The right (sales) people would apply well-oiled processes and demonstrate physical evidence (though free trials, previews, etc). In viral marketing, the sales people can be the VAS users themselves.

Some of these Ps would certainly influence technical solutions, and as such, VAS engineering and marketing would march hand in hand.

This entire approach, however, remains end user centric. According to such approach, very heteroclitic* VAS could be developed at any cost, well in advance of demand, and independently of legacy infrastructure and services — as long as they would be well targeted at certain market segments.

To my opinion, there are multiple potential stakeholders in the success of a VAS, besides residential, business and corporate communication consumers:

> The communication service providers themselves, who need to capitalize on past investments, that is, in PSTN or GSM networks; as such, the new VAS should span different networks by nature
>
> Application service providers, who are going to compare the NGN version of a VAS to the non-NGN (Internet-only) version — IM/VoIP providers, for example, are today not the demanding party for NGN standardization
>
> Regulators and governments, who are looking at widely adopted services but lack of control — are lawful interception, emergency calling, carrier selection and number portability going to be supported for these new communication means?
>
> Advertisers, a party which has propelled many Internet service business models, but seems to be absent from today's telecom services
>
> Device manufacturers, who are sometimes not convinced of the need to use communication rather than storage and computing (remember our discussions in Sections 1.2 and 1.3)
>
> Communities, who need to benefit from these VAS, sometimes even against the interest of individual consumers — for example, the community moderator should be able to refuse new members, scan content, and block end users.
>
> The telecom/IT sector in general, which cannot support another wave of failing hypes after the mobile data bubble in the early 2000s

We could even extend the list of stakeholders to the investors, workforce, and suppliers of these organizations.

Many of these parties deserve to be treated as customers, but they haven't been in the past, in many examples.

---

* Different and incoherent.

Voice-over-IP telephones are now supposed to have stickers with "do not use in case of emergency." Before the TISPAN standards, it is indeed impossible for VoIP service providers to physically locate them. And anyway most VoIP phones, IP routers and WiFi access points do not have a battery, for cases of power outage.

Device manufacturers are facing a fragmented market of form factors (e.g., desktop, laptop, pocket PC, navigator, mobile phone, portable game console, media player). Moreover, they have to select and maintain mobile voice and data communication technologies and provide on-board features such as cameras, Bluetooth, WiFi, firewall, music, games and video players. Maybe we should requalify the manufacturers as device assemblers, because that's what they have become by using so many advanced shared components. They get little indications from network vendors and service providers when it comes to predicting how end users will use the device to communicate.

In a recent survey by the GSM Association, American mobile phone users ranked mobile e-mail as the number one mobile data service, but the GSM service providers, network vendors and 3GPP never specified any enhancement for e-mail delivery. One could have imagined standards for compression, free header previews, lawful interception possibilities, personal storage space for attachments in the network rather than on the mobile device, and so on. Instead, 3GPP and GSMA developed and promoted MMS, a competing technology with per-message charging model.

Classic mobile telephony and text messaging has offered very few opportunities for sponsoring and advertising, contrary to instant messaging clients.

Communities (e.g., my son's soccer club) cannot subscribe to any communication service at all; if a match is canceled, the trainer uses his personal prepaid credit to notify the parents by SMS.

This new, expanded environment is disruptive to classic telcos and cellcos; it goes along the trend to decouple the device, connectivity and service we discussed in Section 1.6.

It requires a transition of the same magnitude as the step from state-owned PTTs (Post, Telecom and Telegraph) to a competitive landscape in the 1990s.

As of now, no single player can pretend to offer the dominating device, connectivity or service anymore, certainly not for a long time. And due to the merging and acquisition activity in the background, it becomes difficult to identify that "single player" anyway. The game is about specialization, complementarity, and collaboration, rather than overlaps and head-to-head competition.

So in summary of this introduction, VAS proposals should be evaluated along an extended list of attributes: the basic forces we discussed in Chapter 1, the 7-P service marketing mix and of course the interests of an extended list of stakeholders.

With this in mind, let's examine a few examples of these VAS.

7.1 Media Pilot

Mobile service providers and device manufacturers have been struggling for a long while to bring the Internet browsing or TV experience to a handheld device, and they still do. The network infrastructure is capable of delivering these services, but the relatively tiny screens on cell phones and absence of keyboard remove the pleasure from browsing. And a convincing digital TV experience starts with fast channel flicking.

The idea here is to start using the NGN device (e.g., IP phone, SIP softphone) as a controller for the media experience (browsing or digital TV). These devices are very likely to include a keypad, so let's try and use that one to control which link is clicked on a Web page, whether the movie needs to play, pause, fast forward, the movements of a character in an online game, and so forth.

With circuit-switched telephony it would have been inconceivable to set up a call and pass DTMF digits (the tones corresponding to key strokes) in-band to the TV station — let alone that the TV station would in those days have been able to provide video or games on demand to individuals.

NGN technology has removed both barriers. The new economics of NGN networks imply that sessions are allowed to last much longer than circuit-switched calls used to be. An NGN user agent has four methods at its disposal to pass DTMF information to the network (we listed them in Section 4.7.7); whether the keypad strokes travel as SIP signaling (INFO, NOTIFY) or in RTP packets doesn't matter (at first sight).

Second, the NGN allows emitting personalized video and audio streams to the end user, using standard encoding/decoding mechanisms. The conversion of Web pages, analog, digital TV channels and games to H.263 and H.324M encoded video, plus the injection of, for example, a pointing arrow in the picture is an important, resource-intensive task.

So it becomes possible for the media pilot, this new NGN VAS provider, to collect these upstream keystrokes, look up the originating user's profile, and pilot the downstream media experience accordingly. A media pilot thus browses the Web, flicks the TV channels or inputs moves to an online game on behalf of the end user. It transforms the results (highlighted Web links, the next Web page, a paused movie, a personal settings menu, etc.) to the device chosen by the end user as NGN media streams or even digital TV.

Pressing "1" on the keypad means clicking that link on the top left corner of the Web page. Pressing "2" means moving the pointing arrow northbound. Pressing a "5" in the middle of the keypad is used to pause or resume a movie on demand. Pressing a "4" means the snowboarder moves to the left in the game. Pressing "6" could make a home surveillance camera turn right. Pressing "8" while navigating Google Earth on your TV will make the image tilt so you can see the horizon.

It is up to the NGN service provider to ensure the upstream and downstream information is delivered with minimal delay.

How far will the pointing arrow move, and how sharp will that snowboarder turn to the left? Well, if the upstream DTMF information remains in band to the NGNSP, or is encoded as in RFC 2833 (including duration), or if the DTMF signals are carried in-band of a circuit-switched call, the NGNSP is able to communicate the duration to the Media Pilot.

Hence a Media Pilot transforms your plain vanilla SIP phone into a universal, standard remote control for everyday life: it will allow you to key in the floor you want to go to, in advance of you entering the elevator. It could allow construction workers to move the crane on a building site. To lock your car's doors, open your garage door, pilot your domotics system, your media center, or the music played at the club: the possibilities are thus infinite and unexplored to date.

Network-based media pilots have the potential to make Windows Vista look like grandma's user interface.

Many applications will arise from the fact that today only one or two remote controls are provided with any appliance, that these remotes have a short range, and are incompatible.

Perhaps this will also call for a motion control protocol to be invented between the media pilot and the end device or application.

7.2 Personal Contact Pages

Personal welcome pages are the evolution of the personalized ring back tone, personalized welcome video, company Web site, and personal Web space. It builds further on the idea from the previous section to turn multimedia user experiences into NGN media streams.

Users of the service receive a default "contact page" which will be played as video, for all session establishment requests (SIP INVITE) arriving at their current telephone number and preferred domain name (e.g., sip:32472700855@mycompany.com). The scope could be extended to other SIP methods such as SIP MESSAGE, or other incoming communications such as traditional circuit-switched calls or texts (SMS).

The provider of this new VAS, the personalized contact page service provider (PCPSP), will thus become the default SIP server for the mycompany.com domain, which would need to be communicated to the public DNS infrastructure. De facto, the PCPSP thus becomes the inbound communication handler (the ICH of Section 4.5).

The idea of a PCPSP of course also responds to the requirement to deliver a contextual portion for NGN communications (Section 1.5).

It is probably not up to the NGNSP to take on this contextual portion; a whole ecosystem of specialized PCPSP might appear in the residential service.

In response to the session establishment request, the originator would be presented an audio stream, video stream or message response, showing information as defined by the destination (called) PCP user. Personalized ring tones, videos, and messages could be returned to the originator, based on the PCP user's preferences (a different ring tone for my parents than for my colleagues) or originating terminal's capabilities. It should be possible to present an existing Web site to the originator (e.g., flash intro of the company) as the video response.

PCP's fundamental objectives are to protect the end user from intrusiveness; therefore a rules-based system would be able to black-list certain originators and take into account the time of week or day of year, or even presence and location events.

It is unacceptable to expect the basic PCP user to go through 10 pages of self care to define the settings, media and rules — even though also that should be accessible for the *aficionados*. The PCPSP should learn this from the past behavior (e.g., calls/messages received, ring tones downloaded, billing relationship, etc) to predict the best settings. In the enterprise environment, some PCP settings would be company-wide, administered centrally, and so on.

A good PCP would be coupled to existing presence and availability information at IM providers, in Lotus/Exchange Calendars, or indicated in real time by PBX attendants.

If the PCP user does not accept the call, video session or message, the PCP SP should propose the originator to store voice, video or text messages.

In the future, PCP would present the originator with buttons to leave an e-mail, contact another responsible via SIP redirection, or immediately leave a message without disturbing the destination PCP user.

A company PCP would hold routing and hunting logic comparable to today's freephone numbers.

From a marketing perspective, the service could be sold to residential, SOHO businesses and corporate customers. It could be given for free to residential users, as there's incoming interconnection revenue, when the NGNSP and PCPSP receive calls and texts from traditional networks.

7.3 Multichannel Media

A frequent complaint about VoIP is the disappointing audio quality (the Mean Opinion Score, we discussed that back in Section 3.6.4).

Hence the idea is to multiply the parallel audio channels transmitted over the NGN, to obtain stereo, 5.1, or even "N.1" sound.

This calls for WiFi microphones that can be positioned in the four corners of your living room and communicate via SIP/RTP, through the NGNSP, to a Multi-Channel Media Service Provider (MCMSP). On the other side, N channels are delivered to the recipient's Bluetooth headset.

The audio stream being played to the recipient's headphones depends on his position relative to the microphones in the room. In other words, in a 5.1 system, as the listener (destination user) moves closer to the back left microphone, the

MCMSP presents a mix of audio streams containing more sound recorded at the back left microphone of the originator.

Let's try and imagine the video version of this service: four SIP/WiFi Webcams record an image on the originating side (front left/right, back left/right) and the image the recipient sees on his portable NGN device (e.g., a portable WiFi-enabled gaming console) depends on the position and orientation of that device relative to his/her own cameras.

The MCMSP calculates a resulting image based on this info, and the four video feeds from the originating cameras.

Alternatively, the image could depend on motion sensors in the recipient device, or Galileo,* instead of external position beacons.

Whatever positioning technology is used, you could explore every angle of a 3D object (a statue on the originating museum for example), by physically walking "around it" on the destination side (your living room).

Now combine video and audio and the communication experience could be tremendous, even with relatively compact handheld NGN devices (screens) on the receiving end.

Multichannel media have the potential to introduce new forms of clubbing/partying, casino, games and sports. Imagine virtually standing on the soccer field during the match, watching the action from a couple of meters, including the audio!

7.4   Combinatory Services

The separation of audio device (microphone and speaker in a Bluetooth headset) and terminal device (camera phone, even already with two cameras) has opened new possibilities to share pictures or video during the voice conversation.

Contrary to the multichannel media above, combinatory services do not require the different media streams to be mixed together to produce a new media stream. As such, they don't rely on massive computing power to process the streams.

It would probably also be fine to launch the service without synchronizing the audio and video streams — the recipient would see the originator's lips moving out of sync, but if this service is meant to be used to film or picture landscapes and interiors, who cares?

Should these multiple media streams then be added to a single session, or may they be negotiated and carried in different sessions? At first sight, it seems fine to do the latter.

Is there a need to have these multiple sessions being established by a single, 500€ phone? Or could I use a 100€ SIP camera and my 100€ cell phone?

---

* The future European GPS system, which will provide accuracy down to a few centimeters, with a subscription.

There are many questions around combinatory services, and it is much too early to rush into (e.g., 3GPP IMS) standardization.

7.5   Contact List Overlay

We have addressed the need for network-based contact lists and tribe memberships in Section 1.11.

It is time for communication service providers to provide the network-based address book, or at least to periodically synchronize the NGN device's address book with the network-based copy.

It's so easy to join a company or soccer club and have access to all employees and members from day one. How much more would we communicate if things were so easy?

How easy would be the decision to change your NGN device monthly, or even have several for different uses, if you didn't have to worry about your address book anymore and, by extension, your agenda, tasks, and messages?

The IETF SIMPLE group has proposed the PIDF (see Section 3.2.3.3), laying the groundwork for a standards-based form of Plaxo, today's leading network-based address book.

Huge developments are to be expected if the NGN User Agent starts to be considered as a contacts repository too, rather than today only as a session establisher, codec negotiator and media encoder/decoder.

7.6   Voice on Web

Is a NGN user agent — a downloadable piece of software as we know it today — installed in your PC for a few months or years?

Since the advent of voice-on-Web,* one can doubt it. In such cases, the software of the user agent is perhaps downloaded and executed as an ActiveX component setting up the session (SIP INVITE). It might even be the site itself (the Web server) registering itself as User Agent in the NGN and setting up the sessions.

It was already obvious from the skins on music playing software, that end users would expect the looks of their music, video and communication software to be fully customizable.

But it would be sufficient to add "NGN buttons" to existing Web sites, "deskbars" to the user's desktop PC environment, and so on.

Perhaps one day we will start a SIP session when clicking on a Google adword, or a term in Wikipedia? Ask the experts …

---

* The abbeyphone.com site by Abbeynet.com is a "SIP VoIP Web portal."

Chapter 8

Future Digital Life

It's a sad evening here in Brussels, November 3, 2043. It's 9: P.M., still quite hot as grandma and I are floating on our water sofa, flicking some membranes, these semi-opaque pictures and videos produced by our brain chips. We're sipping an apple beer and savoring some cashews, remembering the good old times.

This afternoon s7i.bf.eu.bot, our favorite bot, was convicted by DIC — you know, the Digital Intelligence Court — for not paying back her debts.

Your daddy and I had hired excellent lawyer bots to advocate that s7i had accepted the payment scheme proposed by Gtera, its main intelligence and bandwidth provider. But the human prosecutor at DIC, and some neurosites, did not accept the scheme.

They claimed to be first in line for the debt settlement, and that s7i could not liquidate her virtual estate fast enough for them to be paid by noon.

Previously, DIC's inspectors found evidence that s7i had accumulated around 11M cred$ of debt at various neurosites and a few games, sports betting and gambling sites.

I had a session with her banker in the afternoon, but they stopped extending credit lines for bots long time ago.

s7i had become one of my most intelligent bots — among the best 200 or so I have in my collection.

Her digital signature is now going to be deallocated on the public neuronet. A death sentence!

She was created by botfoundry.eu, the bot factory your daddy and I started in 2025.

The NGN had become very dynamic in the mid-2020s. SIP had just defeated HTTP, the historic first king of the Internet, and the NGN had become densely populated with business, government and leisure applications. How could I best describe the atmosphere of that second services revolution, if not perhaps with a poem from Carl Sandburg, "Chicago," from 1916, describing the second industrial revolution that had taken place only 100 years earlier, and the war economy:

> *Hog Butcher for the World,*
> *Tool Maker, Stacker of Wheat,*
> *Player with Railroads and the Nation's Freight Handler;*
> *Stormy, husky, brawling,*
> *City of the Big Shoulders:*
> *They tell me you are wicked and I believe them, for I have seen your painted women under the gas lamps luring the farm boys.*
> *And they tell me you are crooked and I answer: Yes, it is true I have seen the gunman kill and go free to kill again.*
> *And they tell me you are brutal and my reply is: On the faces of women and children I have seen the marks of wanton hunger.*
> *And having answered so I turn once more to those who sneer at this my city, and I give them back the sneer and say to them:*
> *Come and show me another city with lifted head singing so proud to be alive and coarse and strong and cunning.*
> *Flinging magnetic curses amid the toil of piling job on job, here is a tall bold slugger set vivid against the little soft cities;*
> *Fierce as a dog with tongue lapping for action, cunning as a savage pitted against the wilderness,*
> *Bareheaded,*
> *Shoveling,*
> *Wrecking,*
> *Planning,*
> *Building, breaking, rebuilding,*
> *Under the smoke, dust all over his mouth, laughing with white teeth,*
> *Under the terrible burden of destiny laughing as a young man laughs,*
> *Laughing even as an ignorant fighter laughs who has never lost a battle,*
> *Bragging and laughing that under his wrist is the pulse, and under his ribs the heart of the people, Laughing!*
> *Laughing the stormy, husky, brawling laughter of Youth, half-naked, sweating, proud to be Hog Butcher, Tool Maker, Stacker of Wheat, Player with Railroads and Freight Handler to the Nation.*

So were we, in the 2020s, young men building, breaking and rebuilding our bot generation software. And for sure, the Internet was at war. Hell broke loose on the HTTP Web, with terrorist attacks on sites almost every hour. People and businesses having lost everything were trying to rebuild their sites on the SIP NGN.

I had been working in the A&W (avatar and wink) industry in the 2010s, for various kloonie design bureaus. We created the first real video kloonies, with text-to-speech technology, and got their lips moving and their faces being really expressive. We added their very first communication skills, allowing them to act as fully independent NGN user agents. It's also in those days that the Kloonies started to transfer themselves from mobile to mobile, through the SIP session, with their media backpacks full of music and videos.

It's a long story but kloonies themselves were born in 2006, when skype (now neurobay) launched them as simple avatars to cheer up their sessions.

You all know what happened when they connected the first animals (dogs and horses) to the Internet around 2020. One-megasensor Wibree chips had been implanted, and their brain signals would be made available as raw output on the first neurosites. These brain instructions would be received by very simple devices in the front end, like robotic dogs and radio-controlled toy cars. The live animals in the back end would thus watch and hear the front end devices move and speak. As always, the ietf did a great job with their Neuro Session Control Protocol (NSCP, RFC 8682) and Neuro Real Time Protocol (NRTP, RFC 9057) in 2021. The neuronet was born.

A popular joke was that NASA let neurosites control their rovers in the 2022–2024 missions to Ceres. A shame for humanity!

It's your dad who had the idea to connect a kloonie to the neuronet. We mapped the kloonies' SIP protocol and media streams to NSCP and NRTP, as many others had done for other applications. We allowed the kloonies to interact with different neurosites, and to build their own local intelligence. And we called the result "bots."

We saw our creations, under instruction of the neurosites, setting up the sessions and chatting with humans and peer bots. We also allowed them to visit Websites of their choice, watch media, participate in shows and polls, subscribe to rss news feeds, select winks, avatars, and skins, and so on.

The result was strange, and there was nothing the great sessionwall (my Chinese inbound session handler) could do. They DID check my availability, but while I would be fetching you from kindergarten by car, suddenly some crazy bot would invite us for a session, using an awfully ugly avatar barking at us on my car's windshield display. Do you guys remember me shouting, "You swine! Stop draining my fuel cells"?

Probably, in those days, there was too little uplink sensory feedback from the bots, via the neurosites, into the animal brains — therefore preventing the learning effect. The sensor chips were unidirectional, just reading brain signals without giving direct feedback. The animals had to be inside their media barns, to see the video panels and hear the audio feed. And they would be unable to feel, smell, or feel heat, cold, pain and joy.

Unlike braindow's new 3-gigasensor chip, which I had implanted last month in my skull. By the way, its responsiveness is excellent! I can make it do whatever

I want. I just need to think about the old times and I hear my favorite tunes and watch my favorite videos. You should try that new braindow release. If only grandma would upgrade hers, we might get rid of the legacy flat-screen stuff and air-pulsing speakers in our media room. We're already not using the airco anymore; we had sore throats too often.

By 2023, the humans started to confuse our bots with human peers. You couldn't tell the difference anymore, in a short video chat session.

Botfactory's key idea in 2025 was selective breeding. It had been successfully applied to animals and crops as of the nineteenth century, and as of the 2000s, much improved by genetic manipulation.

We created software to evaluate a bot's intelligence, contacts, tribe memberships, financial success, taste for media, online gaming skills, and other qualities. Also, we had assembled an online panel of friendly human reviewers who would give marks to the bot's chat sessions and video behavior.

That allowed us to select the best bots out of each batch, and let them transfer part of their intelligence and media links to the next batch. Probably driven by the animals' natural instincts in the back-end, we would see many and intense sessions between the parent bots and the child bots, much more than with related bots and complete strangers.

We worked in batches of 5 to 6 megabots a month, and kept breeding. We were hosted on Linux server farms that would cost us a million a week (yes only for hosting, database and connectivity!). Luckily some of them accepted to be paid in bots.

And it's with increasing astonishment that we saw more intelligent families of bots appear through the batches. With better musical skills and taste. Or more skilled at online soccer. Or better at war games.

By the late 2020s, things got really out of hand when some companies were reported to have connected inmate brains to the neurosites. They took control over the most aggressive bots, and the broadcasters reported shocking scenes in the online games.

Botfoundry only purchased access to the cleanest, smartest and best managed back-end neurosites (a.o., the Chinese Nao Gan farms). s7i.bf.eu.bot and her sisters were born in 2033 (around 300 but only 20 that were really involved in our company).

The 4G Bot Consortium (4GBC) and Bot Intelligence ASsociation (BIAS) had done an excellent job with their new standards (around 2030).

The NSCP traffic to the neurosites decreased in favor of peer-to-peer communication.

By 2035, when she was aged just 2, grandma, your daddy and I started to have very interesting conversations with s7i.

We would have our daily 20–30 minute video chat, discussing her discoveries and encounters (bots and humans). She would tell me about the news feeds from her tribes, her interesting new contacts, her upcoming calendar, and of course her financial situation in cred$, the digital currency that was introduced in 2023 and is now accepted by billions of applications, neurosites and bots. And by humans of course, in meatspace!

It was a real eye opener for all of us to uncover s7i's ambition and opinions on life, love and health.

Well, you know how she can be. She's a bit like grandma and me. Don't listen too much to her crazy ideas and don't spend too much time at her virtual places. Keep playing real music and doing offline sports. Keep in touch with the real world if you can. It's become a sad world but try and do something good for the off-line people. Their life is already so short compared to ours.

s7i did very well, making 10K cred$ a month — not bad in 2035! She gave apparently interesting multicast sessions, mainly to peer bots of course. No human being could follow the pace at which they would swap terabytes of knowledge files, videochat, or talk with each other.

I stopped scanning her contact lists, and she had become member of far too many tribes. Her interests in virtual clothing and virtual estate brought her far.

A few years ago, famous bots started to create avatars and accessories, inspired by humans, gaming characters, and the media industry.

Some of these early avatars are unique pieces of bot art.

An eBay bot (probably mandated by some museum site) just offered me 200K cred$ for k7i's last avatar: a digitally signed, white-green siren with amazingly human green eyes, long white hair, graceful movements and a soft, charming voice.

You kids and we, the elderly, prefer these bot-designed avatars to your own irl image.

I still remember how the use of telephony numbers faded away around 2025, when neurobay, gtera, and sessionwall swallowed the last ims core networks. People and bots would just dive down to their friends' house on Google Earth, in order to establish their sessions.

The first buildings on Google Earth probably date from the early 2000s, when sketchup dramatically lowered the barrier-of-entry for 3D design.

Later, architects and virtual estate companies appeared and people started to equip their virtual houses with all the comfort we know today. Our street was pretty empty in those days. Road traffic appeared in 2012, when insurance and freight companies looked for a standard way to locate their cars and trucks out there on the road. The first movie on Google Earth was probably only made around 2015. The air transport layer appeared in 2020.

By 2020, the virtual estate market was soaring — your daddy, your uncle, grandma and our bots had purchased land certificates early enough. You wouldn't be living where you are today (I mean, in this beautiful parcel in virtual space of course)!

Today they say that even URIs are going to be swallowed by neurobay's service addressing schemes. It's supported by my braindow chip. I only need to think about my bank, and neurobay Profiles takes me there. Could you imagine typing http://... , your bank's name, ".eu", and clicking plenty of buttons? Totally outdated careful, don't try it while driving!

Around five years ago, in 2038, as many bots, s7i was investing her cred$ savings on the neuronet stock exchange, in energy funds, but also in the virtual estate market of Finland. The real Finland (in meatspace) had become a popular summer beach resort for human tourists from all over the planet, with A390s and B797s flying in millions of tourists every summer, mainly from India and China. But the Finland in neurospace (on Google Earth of course) also seemed to attract hordes of bots visiting it from all over the neuronet. Bots would purchase a virtual apartment over there. It's always been a mystery to me. Would the bots somehow be attracted by Nokia's premises (the historic mobile phone manufacturer)?

s7i also specialized in bot tours, you know, these guided tours in neurospace, mainly up north virtual Scandinavia, where she would guide the tourist bots to interesting virtual places, for a few hundred cred$. Well, I'm stupid — I forgot you went there too.

Three years ago, s7i ended up being hosted on the most prestigious neuronet servers, with numerous redundant instances, excellent bandwidth and great QoS to the neurosites and peer bots.

We humans knew since the 2010s that bots and the neuronet were going to be an amazing universe, but nobody would have thought that human-style behavior could appear so quickly among software bots.

Some of them have of course become worldwide celebrities; I guess your favorites must be craft and 50sent in the com.bot domain, but you know we have rising talent in the company too (alison.bf.eu.bot just to name one).

Tonight I see a live session by mj5.us.bot, in which he is imitating Michael Jackson, the twentieth-century singer/songwriter, and quite honestly, I look at him from all angles and am quite surprised by the realism — nothing to do with the quite artificial image he had on our old DVDs.

I'm sorry that I posted this old lengthy format and that we didn't set up a session tonight with you, but your grandma and I are tired and we're going to bed early. And I know how long our sessions can become, you're such a great girl and there's so much we can learn from you!

Tomorrow we're going to visit the Olympus on Mars — might sound boring to you, but you know, when we were young, we had been told it would only be reachable in real life ...

Acronyms

3G	3rd Generation mobile telephony
3GPP	3rd Generation Partnership Project
ACM	Association for Computing Machinery
ADSL	Asymmetric Digital Subscriber Line
AGCF	(TISPAN PES) Access Gateway Control Function
AGW	(TISPAN PES) Access GateWay
ALG	Application-Level Gateway
AMR	Adaptive Multi Rate
AMR-WB	AMR Wide Band
ANSI	American National Standards Institute
A-RACF	(TISPAN NGN) Access Resource and Admission Control Function
AS	Application Server
ASP	Application Service Provider
B2BUA	(SIP) Back-to-back User Agent
BER	Bit Error Rate
BGCF	(3GPP IMS) Breakout Gateway Control Function
BGF	(TISPAN NGN) Border Gateway Function
BGP	(IETF RFC 4271) Border Gateway Protocol
BHCA	Busy Hour Call Attempts
BICC	Bearer Independent Call Control
BRAS	Broadband Remote Access Server
BS	(IEEE WiMAX) Base Station
CAMEL	(3GPP GSM) Customized Applications of Mobile Enhanced Logic
CAP	(3GPP GSM) CAMEL Application Part
CAPS	Call Attempts Per Second
C-BGF	(TISPAN NGN) Core Border Gateway Function
CDMA	Code Division Multiple Access

CDMA2000	(3GPP2) 3G packet data using CDMA
CDR	Call Data Record
CHAM	(3GPP OSA) Charging Control & Account Management
CI	(3GPP GSM) Cell Identity
CMIP	(ITU-T X.700) Common Management Information Protocol
CMTS	Cable Modem Termination System
COPS	(IETF RFC 2748) Common Open Policy Service
CORBA	(OMG) Common Object Request Broker Architecture
CPE	(eSG) Control Plan Editor
CPU	Central Processing Unit
CSAPS	Call or SMS Attempts Per Second
CSCF	(3GPP IMS) Call Session Control Function
CSLAM	Combined Subscriber Line Access Multiplexer
CSP	Communication Service Provider
DIAMETER	(IETF RFC 3588) Evolution of RADIUS
DNS	(IETF RFC 1035) Domain Name Service
DSCP	(IETF RFC 2474) DiffServ Code Point
DSL	Digital Subscriber Line
DSLAM	Digital Subscriber Line Access Multiplexer
DVB	(ETSI) Digital Video Broadcasting
EAP	(IETF 3748) Extensible Authentication Protocol
EDGE	(3GPP) Enhanced Data rates for GSM Evolution
ENUM	(IETF RFC 3761) tElepnone NUmber Mapping
ETSI	European Telecommunications Standards Institute
FDD	Frequency Division Duplex
FMC	Fixed-Mobile Convergence
FW	FireWall
GERAN	(3GPP GPRS) GPRS Enhanced Radio Access Network
GGSN	(3GPP GPRS) Gateway GPRS Support Node
GPRS	(3GPP) General Packet Radio Service
GSM	(3GPP) Global System for Mobile communications
GTP	(3GPP 09.60) GPRS Tunneling Protocol
HA	(3GPP2 CDMA2000) Home Agent
HA	High Availability
HLR	(3GPP GSM) Home Location Register
HSDPA	(3GPP) High Speed Downlink Packet Access
HSPA	(3GPP) High Speed Data Access (HSDPA + HSUPA)
HSS	(3GPP IMS) Home Subscriber Server
HSUPA	(3GPP) High Speed Uplink Packet Access
I/O	Input/Output
I-BCF	(TISPAN NGN) Interconnection Border Control Function
I-BGF	(TISPAN NGN) Interconnection Border Gateway Function
IEEE	Institute of Electrical and Electronics Engineers

IETF	Internet Engineering Task Force
IM-MGW	(3GPP IMS) IP Multimedia Media GateWay
IMPS	(OMA) Instant Messaging & Presence Protocol
IMS	(3GPP) IP Multimedia Subsystem
IN	(ITU-T Q.12XX) Intelligent Network
INAP	(SS7) Intelligent Network Application Part
IP	(IETF) Internet Protocol
ISC	(3GPP IMS) IMS Service Control interface
ISDN	(ITU-T) Integrated Services Digital Network
ISIM	(3GPP) IMS Subscriber Identity Module
ISO	International Standards Organization
ISUP	(ITU-T SS7) ISDN User Part
ITU-T	International Telecommunications Union, Telecommunications standardization sector
IUA	(IETF SIGTRAN) ISDN User Adaptation
L2TF	(TISPAN NGN) Layer 2 Termination function (at IP Edge)
LA	(3GPP GSM) Location Area
LAN	Local Area Network
LDAP	(IETF RFC 4510) Lightweight Directory Access Protocol
M2PA	(IETF SIGTRAN) MTP2 Peer Adaptation
M3UA	(IETF SIGTRAN) MTP3 User Adaptation
MAN	Metropolitan Area Network
MGC	Media Gateway Controller
MGCF	(3GPP IMS) Media Gateway Control Function
MGCP	(IETF RFC 3435) Media Gateway Control Protocol
MGW	Media GateWay
MGWF	Media GateWay Function
MIB	Management Information Base
MIP	(IETF RFC 3344) Mobile IP
MOML	(IETF) Media Objects Markup Language
MOS	Mean Opinion Score
MML	Man-Machine Language
MMS	(OMA) Multimedia Messaging Service
MMSC	(OMA) MMS Center
MPSP	Media Pilot Service Provider
MRF	(3GPP IMS) Media Resource Function
MRFC	(3GPP IMS) Media Resource Function Controller
MRFP	(3GPP IMS) Media Resource Function Point
MSC	(3GPP GSM) Mobile Switching Center
MSML	(IETF) Media Session Markup Language
MSRP	(IETF SIMPLE) Message Session Relay Protocol
MTP	(ITU-T SS7 Q.70X) Message Transfer Part
MVNO	Mobile Virtual Network Operator

NAPTR	(IETF RFC 3403) Naming Authority Pointer Resource
NASS	(TISPAN NGN) Network Attachment SubSystem
NAT	Network Address Translation
NGN	(IETF) Next Generation Network
NMS	(TMN) Network Management System
NP	Number Portability
OFDM	Orthogonal Frequency Division Multiplexing
OFDMA	Orthogonal Frequency Division Multiple Access
OMA	Open Mobile Alliance
OMG	Object Management Group
ORB	(OMG CORBA) Object Request Broker
OSA	(3GPP) Open Service Access
OTA	Over-The-Air
PAN	Personal Access Network
PBX	Private Branch eXchange
PCF	(3GPP IMS) Policy Control Function
PCPSP	Personal Contact Page Service Provider
PCU	(3GPP GPRS) Packet Control Unit
PDF	(3GPP IMS) Policy Decision Function
PDP	(3GPP GPRS) Packet Data Protocol
PDP	(IETF COPS) Policy Decision Point
PDSN	(3GPP2 CDMA2000) Packet Data Serving Node
PEF	(IETF COPS) Policy Enforcement Point
PES	(TISPAN) PSTN Emulation Subsystem
PKI	Public Key Infrastructure
POTS	Plain Old Telephone Service (delivered by a PSTN)
PPP	(IETF RFC 1661) Point to Point Protocol
PPPoE	(IETF RFC 2516) PPP over Ethernet
PRI	(ITU-T ISDN) Primary Rate Interface
PSTN	Public Switched Telephone Network
QoS	(IETF IP) Quality of Service
RA	(3GPP GPRS) Routing Area
RACS	(TISPAN NGN) Resource & Admission Control Subsystem
RADIUS	(IETF RFC 2864) Remote Authentication Dial-In User Suite
RCEF	(TISPAN NGN) Resource Control Enforcement Function (at IP edge)
RGW	(TISPAN PES) Residential GateWay
RIP	(IETF RFC 2453) Routing Information Protocol
RNC	(3GPP UMTS) Radio Network Controller
RSS	Really Simple Syndication
RSVP	(IETF RFC 2205) Resource ReSerVation Protocol
RTCP	(IETF RFC 3550) Real Time Control Protocol
RTP	(IETF RFC 3550) Real Time Protocol

RTSP	(IETF RFC 2326) Real Time Streaming Protocol
SBC	Session Border Controller
SCCP	(ITU-T SS7 Q.71X) Signaling Connection Control Part
SCCP	Skinny Client Control Protocol
SCF	(3GPP OSA) Service Capability Feature
SCP	(ITU-T IN) Service Control Point
SCS	(3GPP OSA) Service Capability Server
SCTP	(IETF RFC 2960) Stream Control Transmission Protocol
SDP	(IETF RFC 2327) Session Description Protocol
SGSN	(3GPP GPRS) Serving GPRS Support Node
SIGTRAN	(IETF) Signaling TRANsport
SIM	(3GPP GSM) Subscriber Identity Module
SIMPLE	(IETF) SIP for Instant Messaging and Presence Leveraging Extensions
SIP	(IETF) Session Initiation Protocol (RFC 3261)
SMS	(3GPP) Short Message Service
SMSC	(3GPP) SMS Center (SMS Interworking and Gateway Center)
SNMP	(IETF) Simple Network Management Protocol
SOAP	(W3C) Simple Object Access Protocol
SP	Service Provider
SPDF	(TISPAN NGN) Service Policy Decision Function
SS	(IEEE WiMAX) Subscriber Station
SS7	(ITU-T) Signaling System No. 7
SUA	(IETF SIGTRAN) SCCP User Adaptation
TCAP	(ITU-T SS7 Q.77X) Transaction Capabilities Application Part
TCP	(IETF RFC 793) Transmission Control Protocol
TDD	Time Division Duplex
TDM	Time Division Multiplexing
TGCF	Trunking Gateway Control Function (= a MGCF)
TGW	(TISPAN PES) Trunking GateWay (= a MGW)
TIA	Telecommunications Industry Association
TISPAN	Telecoms and Internet converged Services and Protocols for Advanced Networks
TLS	(IETF RFC 4346) Transport Layer Security
TPS	Transactions Per Second
TRIP	(IETF RFC 3219) Telephony Routing over IP
UA	(IETF SIP) User Agent
UDP	(IETF RFC 768) User Datagram Protocol
UE	(3GPP IMS) User Entity
UMA	(3GPP) Unlicensed Mobile Access
UML	(OMG) Unified Modeling Language
UMTS	(3GPP) Universal Mobile Telecommunications System
USIM	(3GPP) Universal Subscriber Identity Module

USSD	(3GPP GSM) Unstructured Supplementary Service Data
UTRAN	(3GPP UMTS) UMTS Terrestrial Radio Access Network
V5UA	(IETF SIGTRAN) V5 interface User Adaptation
VAS	Value-Added Service
VPN	Virtual Private Network
WIN	(TIA IS-826) Wireless IN
WPAN	Wireless PAN
XCAP	(IETF SIMPLE) XML Configuration Access Protocol
XML	(W3C) eXtensible Markup Language
XMPP	(IETF RFC 3920-3923) eXtensible Messaging and Presence Protocol
W3C	World Wide Web Consortium
WAN	Wide Area Network
WiFi	(IEEE 802.11) Wireless Fidelity
WiMAX	(IEEE 802.16-2004) Worldwide Interoperability for Microwave Access
WIN	(TIA IS-826) Wireless Intelligent Network

References

1. The Carphone Warehouse Mobile Life Report 2006, http://www.mobilelife2006.co.uk/.
2. Mobile WiMAX: A Performance and Comparative Summary, Doug Gray, September 2006, http://www.wimaxforum.org/news/downloads/Mobile_WiMAX_Performance_and_Comparative_Summary.pdf.
3. *Architectural Considerations for a New Generation of Protocols*, David D. Clark and David L. Tennenhouse, ACM, New York 1990.

References

Index